Wilfried Dechau

Nossa Punt

Tavanasa · Brücken, Landschaft | Bridges, Landscape | Punts, Cuntrada

Scheidegger & Spiess

Dieses Buch wurde ermöglicht durch Beiträge von:
This book was made possible by contributions from:
Alpiq in Tec Schweiz AG | Serafin Beer, Bad-Ragaz | Caprez Bau AG | Bernard Cathomas, Chur | Lydia und Jürg Conzett, Tamins | J. Erni AG, Bauunternehmung | Gesellschaft für Ingenieurbaukunst | Kulturförderung Kanton Graubünden/Swisslos | Verlag Scheidegger & Spiess AG, Zürich | Stiftung Stavros S. Niarchos, Chur

Die Erhaltung der Versell-Brücke ist dem Einsatz und den Beiträgen folgender Organisationen und Institutionen zu verdanken:
Versell's bridge owes its survival and renovation to the commitment and contributions of the following organizations and institutions:
Uniun und Fundaziun Nossa Punt | Bündner Heimatschutz | Baudepartement und Denkmalpflege des Kantons Graubünden | Bundesamt für Strassen (Astra) Bereich Langsamverkehr und historische Verkehrswege | Pro Patria | Sophie und Karl Binding Stiftung | Stiftung Dr. M. O. Winterhalter

Verlag Scheidegger & Spiess AG, Niederdorfstraße 54, CH-8001 Zürich, Schweiz, Switzerland
(www.scheidegger-spiess.ch)

Fotografie, Redaktion · Photography, Editorial work: Wilfried Dechau, Stuttgart (www.wilfried-dechau.de)
Redaktionelle Mitarbeit · Editorial cooperation: Rita Cathomas-Bearth für Nossa Punt (www.nossa-punt.ch)
Übersetzung ins Englische · English translation: Nicholas Grindell, Berlin
Gestaltung · graphic design: Björn Maser, Böblingen (www.minimalist.cn)
Bildbearbeitung · image production: Elke Weber, ctrl-s, Stuttgart (www.ctrl-s.de)
Druck · printed by: Henkel GmbH Druckerei, Stuttgart (www.henkeldruck.de)
Einband · bound by: Spinner, Ottersweier (www.josef-spinner.de)

ISBN 978-3-85881-587-3

Inhalt

Über sieben Brücken
Wilfried Dechau: Viele Brücken führen über den Rhein 4

Brücken, Landschaft
Ursula Baus: Tavanasa und seine Brücken 9

Dorf im Schatten
Peter Egloff: Gewerbefleiß und ein Schreckenstag 54

Scursanida · Abkürzung
Robert Kruker: Verkehr auf alten Wegen 62

La punt
Arno Camenisch: ... ob die Brücke auch halten würde, ... 66

Porträts und Statements 68

Unsere Brücke
Rita Cathomas-Bearth: Der Weg zu Nossa Punt 84

Porträts und Statements 88

Erhalt der Versell-Brücke
Simon Berger: Suche nach neuer Trägerschaft 95

Maillart · Versell · Menn
Jürg Conzett: Die Versell-Brücke und ihre Vorgängerin 96

Instandsetzung
Andrin Herwig: Die Versell-Brücke wurde instandgesetzt 108

Fundaziun Nossa Punt
Sep Cathomas: Gründung und Stiftungszweck 142

Autoren 144

Content

Seven bridges to cross
Wilfried Dechau: Many bridges lead over the Rhine 4

Bridges, Landscape
Ursula Baus: Tavanasa and its bridges 9

Village in the shade
Peter Egloff: A healthy economy and a day of horror 54

Scursanida · Shortcut
Robert Kruker: Travelling the old paths 62

La punt
Arno Camenisch: ... whether the bridge would hold up, ... 66

Portraits and statements 68

Our Bridge
Rita Cathomas-Bearth: The path to Nossa Punt 84

Portraits and statements 88

Saving Versell's bridge
Simon Berger: Looking for a new owner-operator model 95

Maillart · Versell · Menn
Jürg Conzett: Versell's bridge and its predecessors 96

Restoration
Andrin Herwig: Restoring Versell's bridge 108

Fundaziun Nossa Punt
Sep Cathomas: Founding and Purpose 142

Authors 144

Über sieben Brücken | Wilfried Dechau: Viele Brücken führen über den Rhein

Den Rhein haben die Deutschen immer gern als »ihren« Strom bezeichnet, dabei liegen weder die Quelle noch die Mündung in Deutschland. Er speist sich aus den Graubündner Alpen und mündet in Holland in die Nordsee. Tomas Riehle, einer der einfühlsamsten deutschen Architekturfotografen, hat über viele Jahre das ambitionierte Projekt verfolgt, die Rheinbrücken ins Bild zu setzen. Vom nahe der Quelle quer über das Rinnsal gelegten Brett, von Christian Menns eleganter Bogenbrücke bei Tamins bis hin zu Ben van Berkels Erasmusbrug, einer expressiven Schrägseilbrücke nahe der Mündung in Rotterdam, hat Tomas Riehle 117 Brücken fotografiert. Das sind sehr viele, aber längst nicht alle. Es ging ihm nicht um Vollständigkeit, sondern um die ganze Bandbreite der Brückenbaukunst entlang eines Flusses. Tomas Riehle ist 2017 viel zu früh gestorben, aber das Rhein-Projekt konnte er noch vollenden. Ihm ist dieses Buch gewidmet.

Schade, dass er Jürg Conzetts Punt da Suransuns nicht fotografiert hat. Sie fehlt in der Reihe. Aber Tavanasa ist in Riehles Buch mit zwei Brücken vertreten. Mit der filigranen Stahl-Fachwerkbrücke bei Mutteins und der Bogenbrücke, die Walter Versell 1928 gebaut hat. 2017 wurde diese Brücke saniert. Seither dient sie als Fußgänger- und Fahrradbrücke. Die Instandsetzung habe ich mit der Kamera begleitet und nicht nur die »Golden Gate von Tavanasa« (Marlis Caduff), sondern auch all die anderen Brücken im Kontext ihrer Landschaft fotografiert. Damit nicht genug. Die Rettung der Versell-Brücke gab den Anstoß, mich mit der durch Brücken geprägten Identität und Geschichte des Ortes zu befassen. Ich habe Menschen in Tavanasa kennengelernt, für dieses Buch porträtiert und dazu befragt, wie sie zu dem »Bindeglied zwischen Tavanasa und Danis« (Marc Cathomen) stehen.

Tavanasa. Die Fachwerkbrücke bei Mutteins (links) und Versell-Brücke (rechts).
Die Fotos wurden Tomas Riehles Buch ›Rheinbrücken‹ entnommen (Edition Axel Menges, 2015). Das Buch enthält einen Essay von Gottfried Knapp und ist damit nicht nur sehens-, sondern auch lesenswert

Tavanasa. Truss bridge at Mutteins (left) and Versell's bridge (right). Photos taken from Tomas Riehle's ›Rheinbrücken‹ (Edition Axel Menges, 2015). With an essay by Gottfried Knapp, the book is both worth seeing and worth reading

Seven bridges to cross | Wilfried Dechau: Many bridges lead over the Rhine

The Germans have always liked to call the Rhine »their« river, even though its source and mouth are in other countries. Its headwaters are in the Alps in Switzerland, and it flows into the North Sea in the Netherlands. Over many years, Tomas Riehle, one of Germany's most insightful architectural photographers, pursued the ambitious project of portraying bridges across the Rhine. From a plank laid across the rivulet close to the source, to Christian Menn's elegant arched bridge at Tamins, through to Ben van Berkel's Erasmusbrug, an expressive cable-stayed bridge close to the river's mouth in Rotterdam, he photographed 117 bridges. That is a great many, but by no means all. He was interested not in completeness but in the full range of bridge-building along the course of a river. Tomas Riehle died in 2017, far too soon, but not before completing his Rhine project. This book is dedicated to him.

He didn't photograph Jürg Conzett's Punt da Suransuns, which is a pity, but Tavanasa is represented in Riehle's book by two bridges: the delicate steel truss bridge at Mutteins and the arched bridge built by Walter Versell in 1928. Since its renovation in 2017, the bridge has been used by cyclists and pedestrians. I followed the restoration work with my camera, photographing not only the »Golden Gate of Tavanasa« (Marlis Caduff) but also all of the other bridges in the context of the surrounding landscape.
Beyond this, the salvaging of Versell's bridge prompted me to look at the local identity and history that have been shaped by bridges. I met people in Tavanasa, took portraits of them for this book, and asked them about their views on the »vital link between Tavanasa and Danis« (Marc Cathomen).

Sieben Brücken von Tavanasa | Seven Bridges of Tavanasa

Brücken, Landschaft | Ursula Baus: Tavanasa und seine Brücken

Auf einem Vorderrheinabschnitt von weniger als zwei Kilometern Länge reiht sich in Tavanasa Brücke an Brücke – sieben sind es, die kontinuierlich ihre Zwecke erfüllen, aber zugleich Facetten einer langen Brückenbaugeschichte offenbaren. Man erinnert sich an Wörlitz, wo Ende des 18. Jahrhunderts in einem Park kleine Fußgängerbrücken quasi als Miniaturensammlung zur damals geläufigen Brücken-Typologie entstanden. In Tavanasa reicht der jetzt noch präsente Brückenbau bis ins frühe 20. Jahrhundert zurück – Eisenfachwerk, Stahlbetonbogen und Balkenträger, gerader oder gekurvter Fahrbahnverlauf: Eine bemerkenswerte Vielfalt bewahrter Typen, die den jeweiligen örtlichen Spezifika angepasst wurden, überrascht an dieser Stelle, weil die Orte Tavanasa, Danis, Dardin und Breil/Brigels nicht von wirtschaftlichem Reichtum, nicht von überwältigendem Wachstum zeugen. Die Brücken in Tavanasa sind deswegen gewiss kein Luxus. Auf den folgenden Seiten werden die sieben Brücken knapp vorgestellt. Die Instandsetzungsschritte für die Versell-Brücke werden separat und ausführlich ab Seite 108 erläutert, hier manifestiert sich die Leistungsfähigkeit des zeitgenössischen Ingenieurbaus auch in der Pflege baukulturellen Erbes exzellent.

Blick vom Crest la Crusch, Breil (links). Das Bild zeigt das Vorderrheintal mit Tavanasa von Norden aus. Der Übersichtsplan (oben) wurde daher analog positioniert, das heißt, er ist nicht – wie sonst üblich – genordet

Bridges, Landscape | Ursula Baus: Tavanasa and its bridges

In Tavanasa, a stretch of the Upper Rhine less than two kilometres long is spanned by bridge after bridge, seven in total, that faithfully serve their purpose while also revealing facets of a long history of bridge-building. One is reminded of the park in Wörlitz where, in the late eighteenth century, small footbridges were installed as a miniature guide to the standard bridge typology of the time. In Tavanasa, the surviving bridges stretch back to the early 20th century – steel truss, concrete arch or plate girder, straight or curved road deck: such a diversity of time-tested types, each adapted to its specific location, comes as a surprise since the villages of Tavanasa, Danis, Dardin and Breil/Brigels are marked by neither wealth nor growth. This means that Tavanasa's bridges are certainly no luxury. All seven bridges are briefly presented on the following pages. The various phases in the renovation of Versell's bridge are explained separately, in more detail, beginning on page 108, illustrating the capacity of contemporary civil engineering for preserving architectural heritage.

View from Crest la Crusch, Breil (left). The picture shows the Upper Rhine Valley with Tavanasa from the north. The overview (top) is aligned with the photograph (i.e. it doesn't follow the north-up convention)

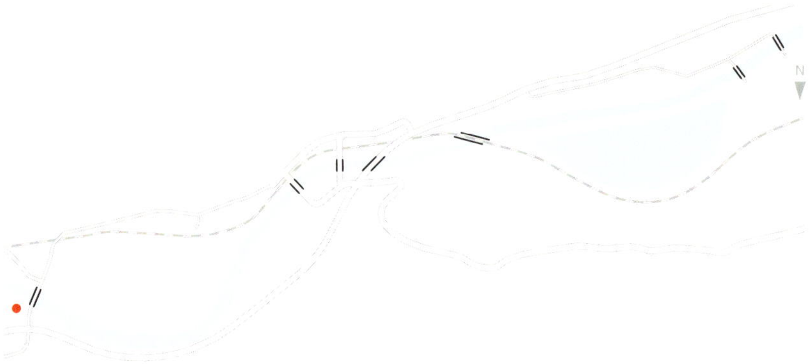

Im Osten Tavanasas führt die Straße aus Richtung Ilanz über eine Stahlfachwerkbrücke in den Ort hinein. 1929 von der Eisenbaufirma Conrad Zschokke mit einer Spannweite von 48,50 Metern gebaut, wurde das federleicht wirkende Bauwerk 2005-06 weitreichend saniert. Es weist eine bemerkenswerte Ausbildung des Übergangs vom Obergurt zur geneigten Endstrebe auf, wobei sich statische und ästhetische Themen überlagern. Bei der Instandsetzung musste die Beton- durch eine orthotrope Stahlfahrbahnplatte ersetzt werden, zudem galt es die Widerlager umzubauen. Korrodierte Stahlprofile wurden ausgetauscht und der Korrosionsschutz erneuert. Gelungen ist es dem Büro Conzett Bronzini Gartmann, den filigranen Charakter der Brücke im Sinne des Ingenieurbaus als Minimal Art zu erhalten.

In the east, the road from Ilanz comes into Tavanasa over a steel truss bridge. Built in 1929 by the company of Conrad Zschokke with a span of 48.50 metres, the airy-looking structure was extensively restored in 2005-06. It features a remarkable transition from the top chord to the inclined end strut in which structural and aesthetic themes overlap. During restoration work, the concrete road deck had to be replaced with one made of orthotropic steel and the abutments were modified. Corroded steel girders were replaced and the anticorrosive coating was renewed. Conzett Bronzini Gartmann succeeded in preserving the bridge's delicate character in the spirit of civil engineering as Minimal Art.

7t
3,7m

VERBINDUNGSSTRASSEN

Kanton Graubünden

Strassennummer

723.00

Strassenzug	Bauabschnitt:	km
TAVANASA INNERORTS		0.00
ZUR OBERALP-STRASSE	VORDERRHEINBRÜCKE VOR TAVANASA	

Effektive Baulänge: m

WERKSTATTPLÄNE 1929

VORDERRHEINBRÜCKE VOR TAVANASA

Hauptträger 1:20

Plan Nr.: 723.00.317.3 Datum: Mai 2006

Kantonale Behörde:	Eingangsstempel	Der Projektverfasser:
Die Regierung des Kantons Graubünden Tiefbauamt Graubünden		Conzett, Bronzini, Gartmann AG dipl. Ingenieure ETH / FH / SIA Bahnhofstrasse 3, CH-7000 Chur Tel. 081/258'30'00 Fax. 081/258'30'01 Email cbg@cbg-ing.ch
	Kunstbautennummer 23 00 05	

Achtung:
Dieser Plan ist nur zum Teil gültig, da die Brücke im Jahre 2006 instandgesetzt wurde.

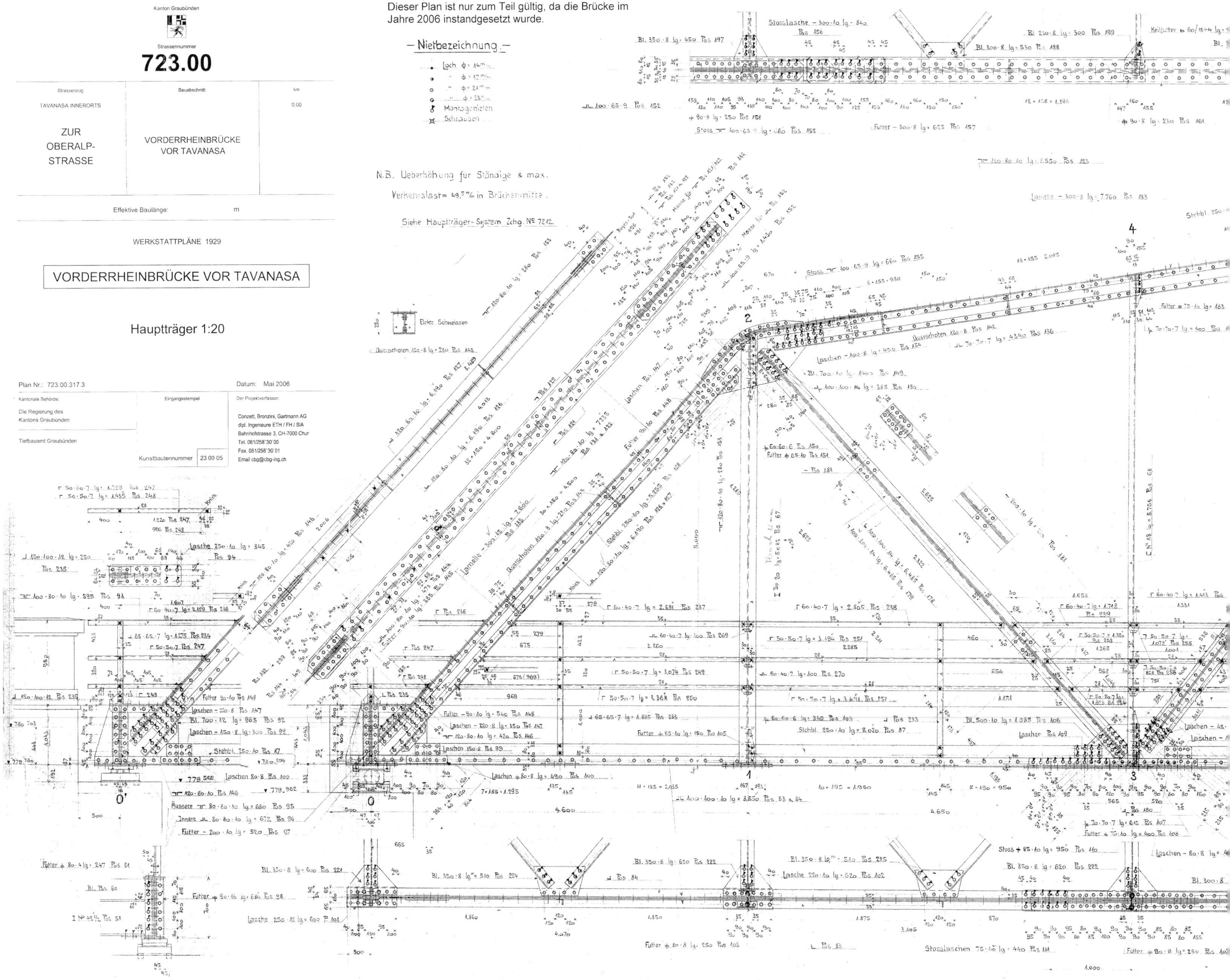

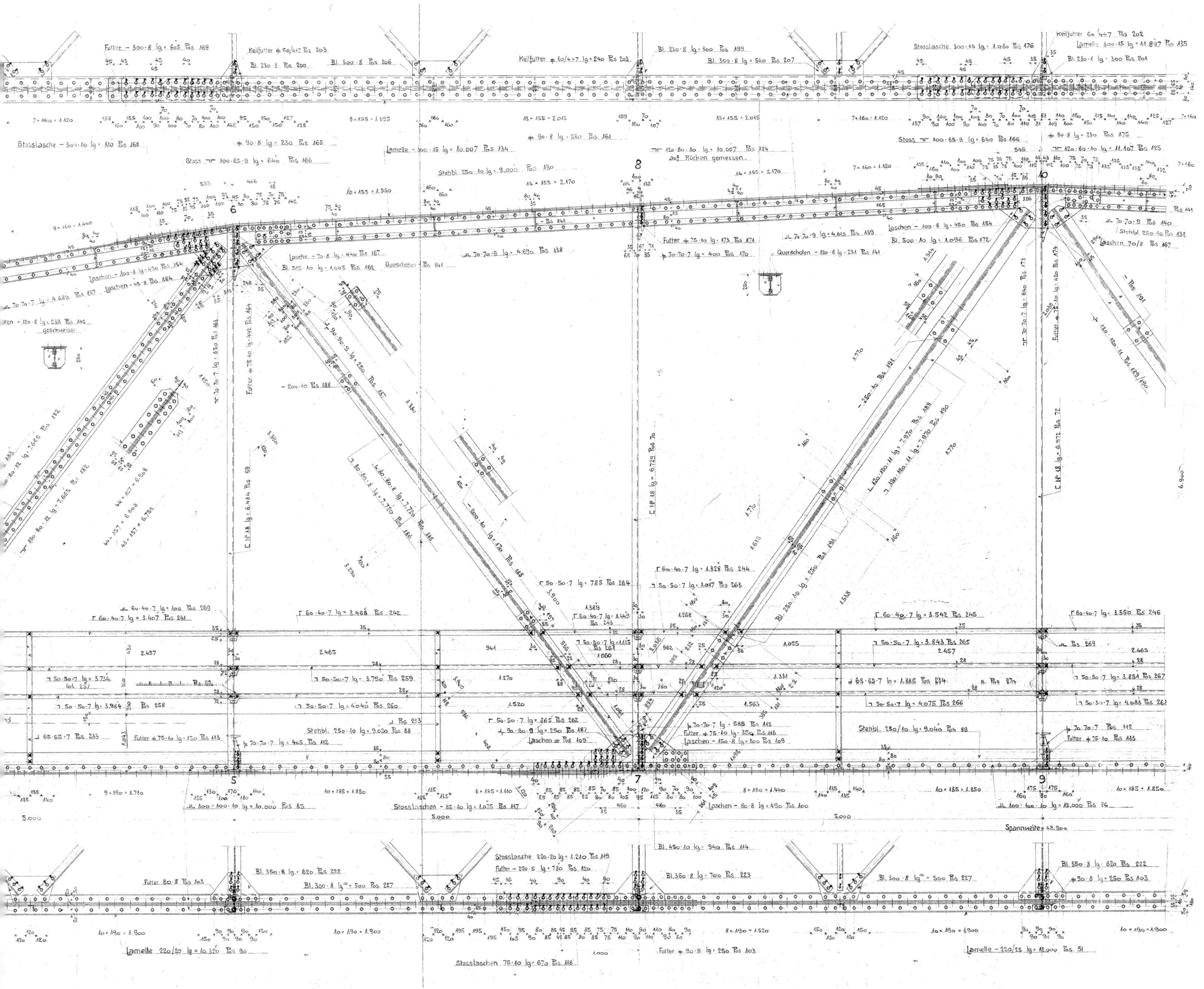

Futter - 300·8 lg = 605 Pos 169
Keilfutter ⌀ 60/4÷7 Pos 203
Bl. 230·8 Pos 200
Bl. 300·8 Pos 206
Keilfutter ⌀ 60/4÷7 lg = 240 Pos 203
Bl. 230·8 lg = 300 Pos 199
Bl. 300·8 lg = 560 Pos 207
Stosslasche 300·15 lg = 1.080 Pos 176
Keilfutter 60/4÷7 Pos 202
Lamelle 300·15 lg = 11.897 Pos 135
Bl. 230·8 lg = 300 Pos 201
Stosslasche - 300·10 lg = 810 Pos 168
⌀ 90·8 lg = 230 Pos 165
Stoss ⊤ 100·65·8 lg = 640 Pos 166
Lamelle - 300·15 lg = 10.007 Pos 134
⌀ 90·8 lg = 230 Pos 161
⊤ 120·80·10 lg = 10.007 Pos 124
auf Rücken gemessen.
Stehbl. 250·10 lg = 9.000 Pos 130
Stoss ⊤ 100·65·8 lg = 640 Pos 166
⌀ 90·8 lg = 230 Pos 175
⊤ 120·80·10 lg = 11.107 Pos 125
Laschen - 100·8 lg = 450 Pos 154
Bl. 500·10 lg = 1.096 Pos 172
⊥ 70·70·9 Pos 140
Stehbl. 250·10 Pos 131
Laschen 70/8 Pos 167
Futter ⌀ 75·10 lg = 173 Pos 171
⊥ 70·70·7 lg = 400 Pos 170
⊥ 70·70·9 lg = 4.613 Pos 139
Querschoten - 120·8 lg = 231 Pos 141
Lasche - 70·8 lg = 440 Pos 167
Bl. 515·10 lg = 1.005 Pos 162
Querschoten Pos 141
⊥ 70·70·9 lg = 4.690 Pos 138
Laschen - 100·8 lg = 430 Pos 154
Laschen - 45·8 Pos 184
⊥ 70·70·7 lg = 4.685 Pos 157
- 120·8 lg = 238 Pos 142
geschweisst
- 200·10 Pos 188
Futter ⌀ 75·10 lg = 442 Pos 164
⊥ 70·70·7 lg = 650 Pos 163
Futter ⌀ 75·10 lg = 640 Pos 173
⊥ 70·70·7 lg = 640 Pos 174
- 120·120·11 Pos 189/190
- 250·10 Pos 191
⊏ NP 18 lg = 6.484 Pos 69
⊏ NP 18 lg = 6.729 Pos 70
⊏ NP 18 lg = 6.972 Pos 72
- 200·10 lg = 170 Pos 188
Bl. 250·10 lg = 850 Pos 191
⌐ 50·50·7 lg = 785 Pos 264
⌐ 60·40·7 lg = 1.328 Pos 244
⌐ 50·50·7 lg = 1.017 Pos 263
⊥ 60·40·7 lg = 100 Pos 269
⌐ 60·40·7 lg = 3.407 Pos 241
⌐ 60·40·7 lg = 3.466 Pos 242
⌐ 60·40·7 lg = 1.449
⌐ 60·40·7 lg = 3.542 Pos 245
⌐ 60·40·7 lg = 3.550 Pos 246
⌐ 50·50·7 lg = 3.843 Pos 265
⌐ 50·50·7 lg = 3.734
⌐ 50·50·7 lg = 3.790 Pos 259
⊥ 65·65·7 lg = 1.886 Pos 234
⌐ 50·50·7 lg = 3.854 Pos 267
⌐ 50·50·7 lg = 3.984 Pos 258
⌐ 50·50·7 lg = 4.040 Pos 260
⌐ 50·50·7 lg = 4.075 Pos 266
⌐ 50·50·7 lg = 4.088 Pos 268
Stehbl. 250·10 lg = 9.020 Pos 88
⌐ 50·50·7 lg = 865 Pos 262
⌀ 90·90·9 lg = 250 Pos 187
Laschen = Pos 109
⊥ 70·70·7 lg = 585 Pos 115
Futter ⌀ 75·10 lg = 350 Pos 116
Laschen - 150·8 lg = 500 Pos 109
Stehbl. 250/10 lg = 9.040 Pos 89
⊥ 70·70·7 Pos 112
Futter ⌀ 75·10 Pos 113
⊥ 65·65·7 Pos 233
Futter ⌀ 75·10 lg = 150 Pos 113
⊥ 70·70·7 lg = 405 Pos 112
⊥ 100·100·10 lg = 10.000 Pos 85
Stosslaschen - 85·10 lg = 1.075 Pos 117
Laschen - 80·8 lg = 490 Pos 100
⊥ 100·100·10 lg = 12.000 Pos 36
Spannweite = 48.500
Bl. 450·10 lg = 940 Pos 114
Stosslasche 220·20 lg = 1.210 Pos 119
Futter - 220·5 lg = 780 Pos 120
Futter 80·8 Pos 103
Bl. 350·8 lg = 620 Pos 222
Bl. 300·8 lg = 500 Pos 227
Bl. 360·8 lg = 700 Pos 223
Bl. 300·8 lg = 500 Pos 227
Bl. 350·8 lg = 620 Pos 222
⌀ 90·8 lg = 250 Pos 103
Lamelle 220/20 lg = 10.320 Pos 90
Futter ⌀ 90·8 lg = 250 Pos 103
Stosslaschen 75·10 lg = 670 Pos 118
Lamelle - 220/25 lg = 12.000 Pos 91
5
6
7
8
9
10

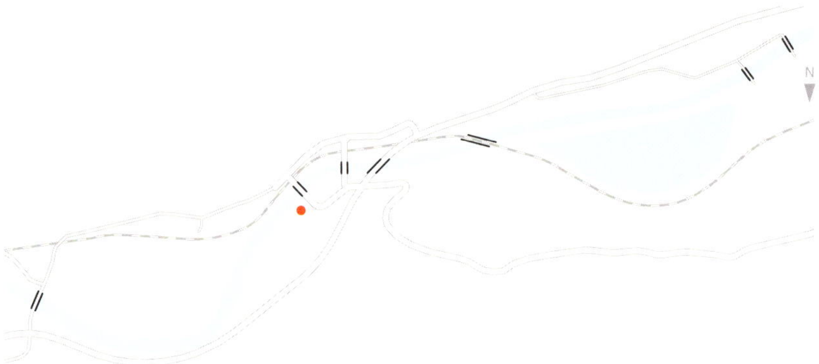

Was in der Schweiz dafür geleistet wird, Landschafts-, Sozial- und Baugeschichte im Sinne eines ideologisch unbelasteten Heimatschutzes, im Sinne bautechnischen Wissens und gestalterischer Kompetenz zu pflegen, sucht weltweit seinesgleichen. Die Vorderrheinquerung, die Walter Versell 1928 zwischen Tavanasa und Danis als Ersatz für eine Brücke von Robert Maillart gebaut hatte, konnte – darum geht es in diesem Buch – gerettet werden. Robert Maillart und Walter Versell: Deren Werke opfert man nicht. Man muss retten, was zu retten ist. Und man muss wissen wie. Dazu lohnt es sich weiterzulesen auf Seite 96 und 108.
Wichtig wird es sein, die Wegenetze von Fußgängern und Radfahrern jetzt auf beiden Seiten des Flusses auf diese wiederhergestellte Flussquerung abzustimmen.

In its unideological efforts to preserve landscapes, social and architectural history in the spirit of engineering expertise and design skill, Switzerland is unparalleled. The bridge over the Upper Rhine built in 1928 by Walter Versell between Tavanasa and Danis to replace a bridge by Robert Maillart has been saved – the subject of this book. Works by Robert Maillart and Walter Versell should not be sacrificed. What can be saved, must be saved. And one has to know how. For more on this subject, continue reading on pages 96 and 108.
Now it will be important to adjust the networks of foot- and cycle-paths on both sides of the river to this restored crossing point.

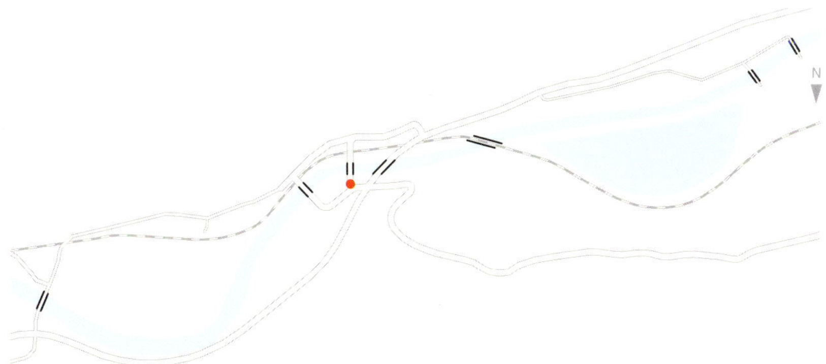

Ursprünglich war geplant, die alte Brücke nach dem Bau der neuen abzureissen. Jetzt bleiben beide, die sanierte Versell-Brücke ist den Fußgängern und Radfahrern vorbehalten, die neue dient vor allem dem Autoverkehr von Tavanasa nach Danis, Dardin und Breil/Brigels. Hier kam es auf einen rücksichtsvollen, ortsbezogenen Brückenentwurf an, weil die Vorderrheinlandschaft reizvoll ist, geradezu eine Attraktion. Die Ingenieurgemeinschaft Conzett Bronzini Partner und Bänziger Partner baute eine vierfeldrige Balkenbrücke aus vorgespanntem Beton und positionierte die Brückenpfeiler den örtlichen Gegebenheiten entsprechend – so ergaben sich die Spannweiten von 16, 26, 36 und 22 Metern. Stützen- und Brückenquerschnitte überzeugen mit eleganter Gestaltung, und raffiniert ist zugleich, wie Werkleitungen und Abdichtungsentwässerung kaum sichtbar integriert worden sind.

Originally, the plan was to demolish the old bridge once the new one was finished. Now both have survived: Versell's bridge is reserved for pedestrians and cyclists, while the new bridge is mainly for motorized traffic from Tavanasa to Danis, Dardin and Breil/Brigels. Here, the design had to take the location into account because the beauty of the Upper Rhine landscape makes it an attraction. The engineering group Conzett Bronzini Partner and Bänzinger Partner built a four-span beam bridge in prestressed concrete, positioning the piers in accordance with the terrain to create spans of 16, 26, 36 and 22 metres. The bridge and its supports are elegantly designed, a further refined touch being the way the cables and drainage systems are integrated so as to be barely visible.

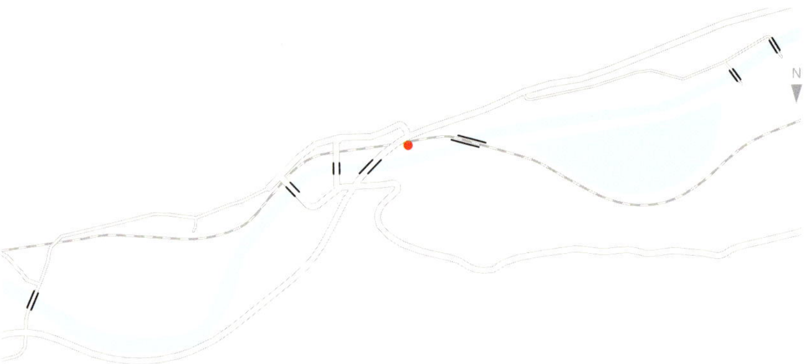

Vom Durchgangsverkehr entlastet war Tavanasa erst, als 1976-77 die Umfahrungsstraße und in dem Zusammenhang eine Brücke gebaut wurden, die sich in einer weiten Kurve (Radius etwa 350 Meter) mit einem Durchlaufträger über den Vorderrhein spannt. Feldweiten von zwei Mal 37 und zwei Mal 61,60 Metern, Hohlkastenprofil und gleichförmige Stützenquerschnitte kennzeichnen diese Vierfeldbrücke, die vom Ingenieurbüro E. Toscano aus Pontresina für den Kanton Graubünden gebaut worden ist. Die Umfahrungsstraße führt hier auch über die Rhätische Eisenbahn und natürlich einen Uferweg. Die Untersicht der Brücke entspricht hier also in etwa der »fünften Fassade« eines Hauses, von der man spricht, wenn die Dachaufsicht gut einsehbar ist.

Tavanasa was not relieved of through traffic until 1976-77 when the bypass was built, including a bridge spanning the Rhine in a broad curve (roughly 350 metres in radius) with a continuous beam. The bridge, built for the Canton of Grisons by civil engineers E. Toscano of Pontresina, has four spans (two of 37 metres and two of 61.60), a box girder and uniform pier sections. At this point, the bypass also passes over the Rhaetian Railway and a riverside footpath. The resulting view of the bridge from below roughly corresponds to the »fifth facade« of a house, a term used when the bird's eye view is readily visible.

65
70
6
520

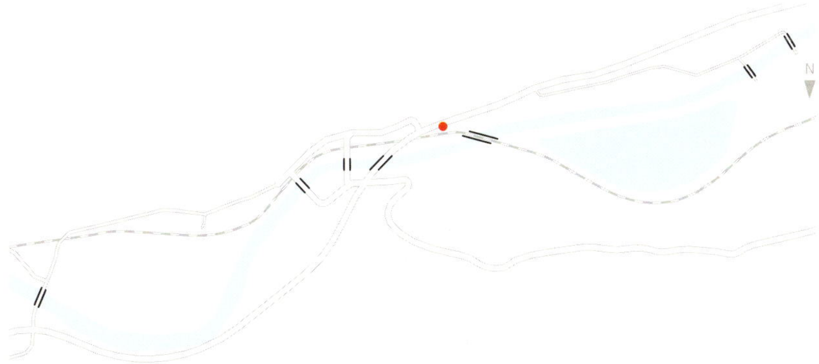

Die Bahnstrecke führt westlich von Tavanasa, unterhalb von Danis, über eine genietete Stahlfachwerkbrücke mit einer Spannweite von 71,50 Metern über den Vorderrhein. In der Übergangskurve stimmt die Gleisachse lokal nicht mit der Brückenachse überein. 1910-11 von der Brückenabteilung der Maschinenfabrik Theodor Bell & Cie. aus Luzern gebaut, weist der elegante Bau – wie die Brücke bei Mutteins – bemerkenswerte Knotenausbildungen auf.
Etwa alle fünfzig Jahre muss bei solchen Bauten der Korrosionsschutz erneuert werden – auch muss man die Lagerungen der Eichenschwellen kontrollieren. Bei der Instandsetzung 2000 wurden deshalb der Schwellenrost erneuert und das ursprüngliche, längs verschiebliche Rollenkipplager durch ein neues Topflager ersetzt.

To the west of Tavanasa, below Danis, the railway passes over the Rhine on a riveted steel truss bridge with a span of 71.50 metres. In the transition curve the path of the track departs from the centre line of the bridge. Built in 1910-11 by the bridge division of the Theodor Bell & Cie. machine factory in Lucerne, this bridge, like the one at Mutteins, features remarkable joint formations.
The anticorrosion coating on such structures must be renewed roughly every fifty years, and the bearing of the oaken sleepers must also be checked. When the bridge was renovated in 2000, the sleepers were renewed and the original longitudinal roller bearing was replaced by a new pot bearing.

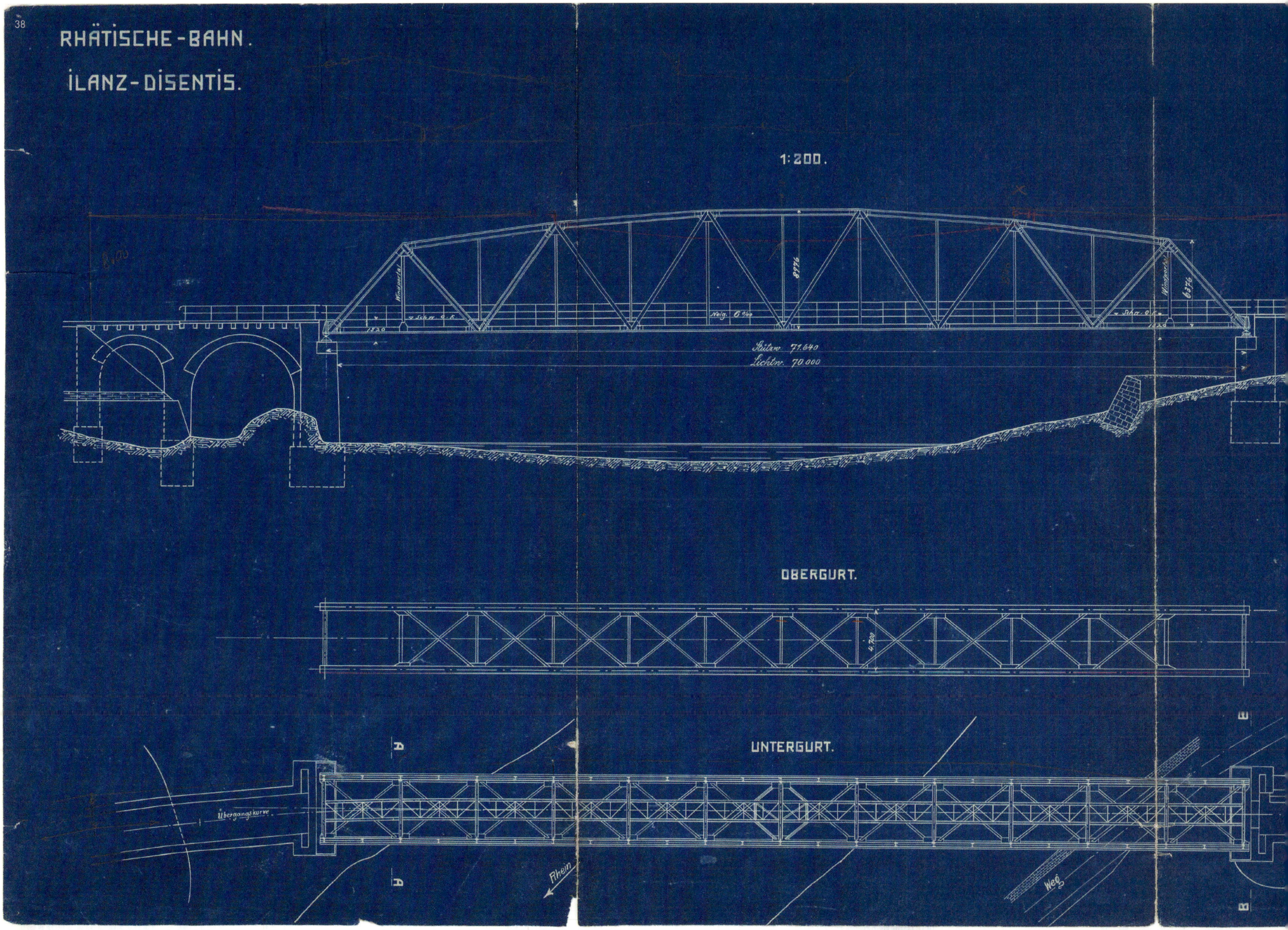
RHÄTISCHE-BAHN.
ILANZ-DISENTIS.
1:200.
Stützw. 71.640
Lichtw. 70.000
OBERGURT.
UNTERGURT.
Übergangskurve.
Rhein
Weg

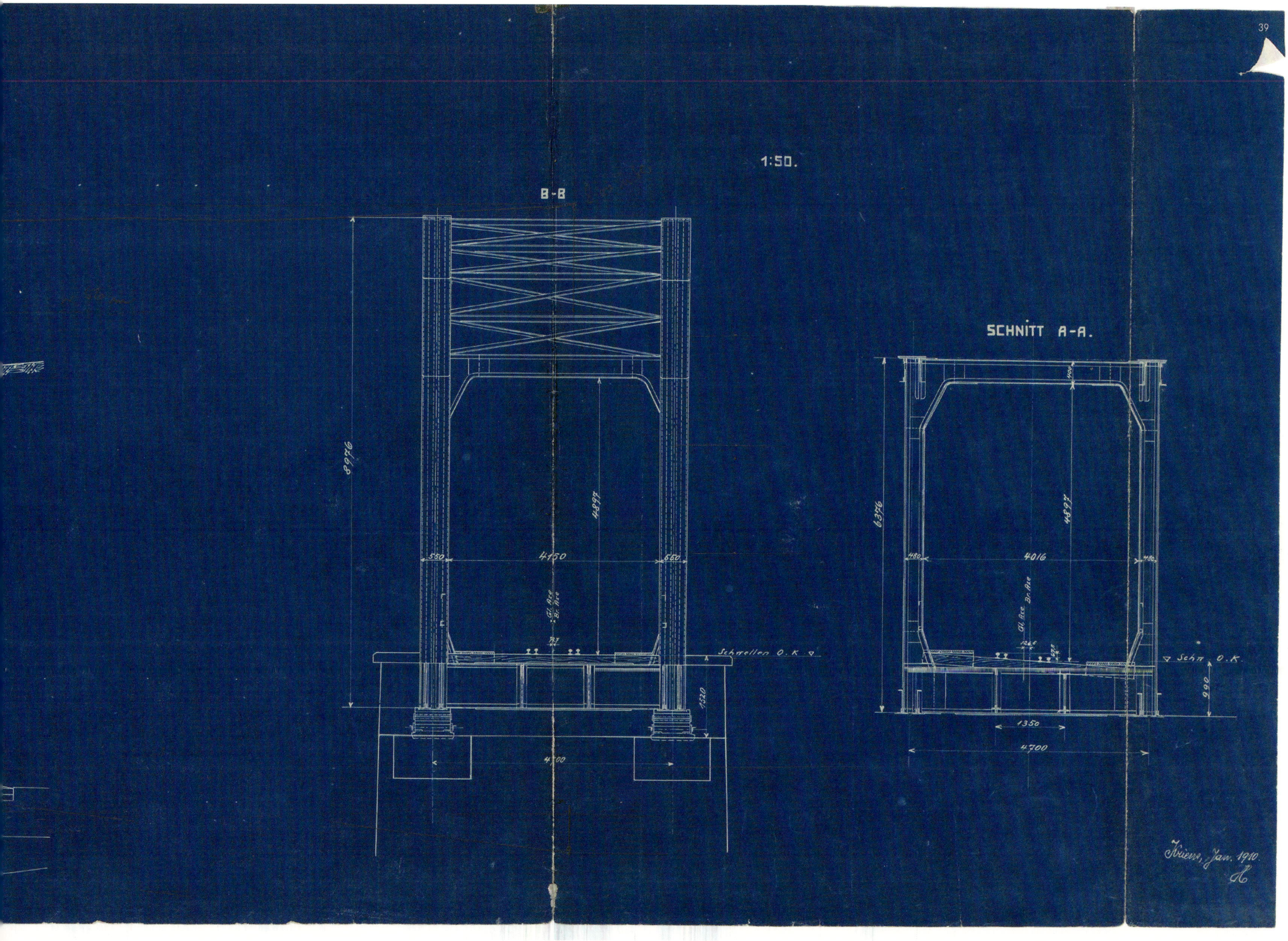
1:50.
B-B
SCHNITT A-A.
8976
4897
550
4150
650
Schwellen O.K.
1320
4700
6376
4897
480
4016
480
Schn O.K.
990
1350
4700
Kriens, Jan. 1910.

65
70

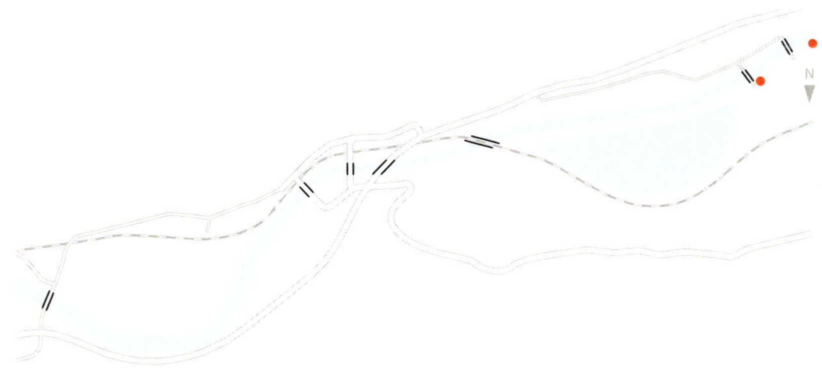

Am Westende des Vorrheinabschnitts von Tavanasa entstand Mitte der 1940er-Jahre ein Wasserkraftwerk, in dessen Kontext Ende der 1950er-Jahre eine Kabelbrücke und eine Wehrbrücke vom Ingenieurbüro Hunger aus Chur errichtet wurden. 1962 ging das Speicherkraftwerk in Betrieb. 1973 setzte das Büro Leo Condrau aus Chur die einfachen, vorgespannten Balkenbrücken instand. Obwohl für den Kraftwerksbetrieb gebaut, sind die beiden Brücken öffentlich zugänglich und leisten Fußgängern und Radfahrern beste Verkehrsdienste.
Die Querungen über den Vorderrhein sind mit der Instandsetzung der Versell-Brücke zu einem guten Abschluss gekommen. Nun möge es gelingen, die Wege auf Tavanasa- und Danis-Seiten stimmig, das heißt auch funktional, anzulegen. Einheimische und Gäste werden es danken.

In the 1940s a hydraulic power station was built at the western end of Tavanasa's stretch of the Upper Rhine and at the end of the 1950s the engineering firm of Hunger from Chur built a cable bridge and a weir bridge to serve it. In 1962 the storage power plant went into operation. In 1973 the firm of Leo Condrau from Chur restored the simple, prestressed beam bridge. Although built for the power stations, both bridges are open to the public, rendering excellent services to pedestrians and cyclists.
With the restoration of Versell's bridge, the Rhine crossings have been happily completed. Hopefully the tracks and footpaths on both sides, in Tavanasa and Danis, can now be functionally coordinated – to the benefit of both local residents and visitors.

Ovras Electricas Reinanteriur SA
axpo
axpo

Ort Tavanasa | The village of Tavanasa

Dorf im Schatten | Peter Egloff: Gewerbefleiß und ein Schreckenstag

Genau jene Stelle im »Lochli«, an der die Mure niedergegangen war, hatten sich die Straßenbauarbeiter ausgewählt, um ihren Abraum zwischenzulagern (September 2017). Peter Egloff machte den Baggerfahrer darauf aufmerksam, was hier geschehen war und dass ein großer Schotterhaufen – neunzig Jahre danach – so gespenstisch wirkt

The construction workers chose the exact spot in the »Lochli« where the landslide had come down to store their waste (September 2017). Peter Egloff explained to the excavator driver what had happened here and why a large pile of gravel – ninety years later – looked so eerie

»Via da batten, via da nozzas, via da bara«, weiß das romanische Sprichwort. Zu Taufe, Hochzeit und Begräbnis führt derselbe Weg, und zweimal wird der Mensch dabei getragen. Tavanasa ist nach Danis kirchgenössig, seit jeher hat der Weg zum Gotteshaus über die Rheinbrücke geführt.

»Dass Salvator Caduff ein geachteter Mitbürger und geschätzter Lehrer war, bewies der lange Trauerzug, der sich am 23. April 1985 von seinem Haus in Tavanasa nach Danis zur Dorfkirche hinauf bewegte«, vermerkt der Nekrolog im »Bündner Schulblatt«. Vierzig Jahre lang war Salvator Caduff Primarschullehrer drüben in Danis. Und für ebenso viele Jahre gottgefälligen Orgelspiels wurde ihm die päpstliche Verdienstmedaille »Bene merenti« verliehen. Zusammen ergibt das viele, viele Tausend Brückengänge.

Salvator Caduff wird an einem frostigen Januartag des Jahres 1913 über Robert Maillarts Brücke zum Taufstein getragen. Er benutzt sie als Kind und Jugendlicher, er wird Zeuge ihres Untergangs, bestaunt den zugigen Bau der Nachfolgerin. Und er zeigt zeitlebens ein waches Interesse für das Werden, Sein und Vergehen in seinem Dorf, das von Anfang November bis Mitte Februar kein Sonnenstrahl erreicht. Wenige Monate vor seinem Tod will Salvator sein Wissen weitergeben und bringt es zu Papier: »Entginas notizias da Danis-Tavanasa« | Einige Notizen zu Danis-Tavanasa. Seine Tochter Mengia hat sie mir überlassen, diese notizias des Lehrers und Organisten Salvator Caduff-Zortea, der jetzt selber schon so manches Jahr dunkle Friedhofserde ist, drüben in Danis.

»Mo purs havevan nus fetg paucs a Tavanasa« | Reine Bauernbetriebe hatten wir sehr wenige in Tavanasa, berichtet Chronist Salvator. Dafür gab es zu wenig nutz-

Village in the shade | Peter Egloff: A healthy economy and a day of horror

»Via da batten, via da nozzas, via da bara«, goes the Romansh saying. The same road leads to one's baptism, wedding and funeral, and on two of these journeys we are carried. Tavanasa is part of the parish of Danis and the path to church has always led over the Rhine bridge. »Salvator Caduff's status as a respected fellow citizen and esteemed teacher was proved by the long funeral procession that wound its way up from his house in Tavanasa to the village church in Danis on 23 April 1985«, noted the obituary in the Bündner Schulblatt. For forty years, Caduff had taught at

Hier war die Werkstatt von Rechenmacher Barclamiu Zortea

Former location of Barclamiu Zortea's rake-making workshop

Rechenmacher Giusep Speschas Eisenwarenhandlung wird von seinen Nachkommen weitergeführt

Rake-maker Giusep Spescha's hardware store is still run by his descendants

Die bis 1954 betriebene Gerberei Caduff

The Caduff Tannery, that operated until 1954

Das Wohnhaus von Lehrer und Dorfchronist Salvator Caduff

House of the teacher and village chronicler Salvator Caduff

Am »Lochli« stand bis 1927 die Glockengießerei Alig

Until 1927, the Alig bell foundry stood at the »Lochli«

Das versetzte Memento für die Toten von 1927

The relocated memorial to those who died in 1927

In der Kuhglocken-Sammlung von Maurus Cathomas finden sich neben zahlreichen Exemplaren von »Alig Tavanasa« und »Moser Danis« auch Glocken mit der Bezeichnung »Alig Moser« oder »Moser Alig«

In his collection, Maurus Cathomas has many cowbells marked »Alig Tavanasa« and »Moser Danis«, but also some with »Alig Moser« or »Moser Alig«

bares Land zwischen steilem Hang und Rhein. Aber: »Nus lessen menziunar che bunamein tuts havevan sper lur mistregn, tgi in pli grond e tgi in pli pign menaschi puril, ina sanadeivla economia.« | Wir möchten erwähnen, dass fast alle neben ihrem Handwerk eine kleinere oder größere Landwirtschaft hatten, eine gesunde Ökonomie. Das spärliche Wies- und Ackerland und damit die Notwendigkeit, neben der Landwirtschaft ein zusätzliches Einkommen zu erzielen, stand am Beginn von Tavanasas Gewerbefleiß, bestimmte seine vielfältige und, wie Salvator fand, gesunde Wirtschaft mit Handwerksbetrieben, die ihre zumeist bäuerliche Kundschaft in der ganzen Region fanden.

Ignaz Derungs, als Internet-Chronist von Danis-Tavanasa in den Fußstapfen von Salvator Caduff unterwegs (www.ils-derungs.ch), hat eine Liste erstellt und ist auf nicht weniger als 31 Gewerbe und Berufe gekommen, die in Danis-Tavanasa ausgeübt wurden und – zum kleineren Teil – noch werden.

»Lein entscheiver entadem Tavanasa« | Lass uns zuhinterst in Tavanasa beginnen, schlägt Salvator Caduff vor. »Leu steva sper la val dil Lochli la casa dalla fam. Alig.« | Dort stand neben der Val dil Lochli das Haus der Familie Alig. Die Aligs sind Zugezogene. Kurz vor Mitte des 19. Jahrhunderts steigt ein Josef Anton Alig von Obersaxen herunter ins schattige Tavanasa, lässt sich am Lochli nieder, lernt Romanisch, nennt sich fortan Giusep Antoni und beginnt, aus Kupfer und Zinn bransinas, Kuhglocken, zu gießen. Sein Sohn, ein Giusep Antoni auch er, heiratet Turtè Riedi aus Danis und führt den väterlichen Betrieb weiter, auch zwei ihrer Söhne erlernen das Glockengießer-Handwerk. Gröbere ethnische Vorurteile scheinen den Aligs fremd zu sein: Gut und gern arbeiten sie gelegentlich mit zwei Jenischen zusammen. Johann Friedrich Moser ist von Obervaz nach Schnaus und später nach Danis gezogen, Moser Danis steht auf den Glocken, aber oft auch Moser Alig oder Alig Moser. Auch Sohn Ferdinand arbeitet mit den Aligs, ist als Fahrender indes viel unterwegs und gießt am liebsten auf der Stör.

»AVIS Der Unterzeichnete gibt bekannt, dass er auch dieses Jahr defekte Kuhglocken repariere, neue herstelle, große und kleine verkaufe, solche gegen Altmetall oder auch Hausrat eintausche. – Wer Arbeiten schickt, soll die genaue Adresse angeben und sagen, wie die Arbeit gewünscht wird. – Zum Disentiser Martinimarkt vom 12. November kommt der Unterzeichnete selber mit einem Posten Kuhglocken, die für sehr vernünftige Preise erhältlich sind. – Die Kuhglocken sind weit herum bekannt und tragen die Inschrift A l i g. Gius. Ant. Alig, Kuhglockengiesser, Tavanasa.« (Inserat in der Gasetta Romontscha, 16.10.1924, Bild rechts)

Wenn sie nicht Glocken gossen, lese ich bei Salvator, löteten Vater Alig und seine Söhne Eimer und Zuber aus Blech – und waren passionierte, geschickte Fischer. Die ersten offenbar in Tavanasa, die neben Netz und Reuse auch Rute und Angel verwendeten. Toni Alig, ein schmächtiges Männchen, habe einst in Dautras, wo sich jetzt seit den 1960er-Jahren die Anlagen der Vorderrhein-Kraftwerke dehnen, eine Seeforelle von 18 Pfund gefangen. Über alles lieb und teuer aber ist Salvator die Erinnerung an die hellen Klänge, wenn er als Bub am Lochli vorüberging und die bransiners eben eine Serie frisch gegossener Glocken kontrollierten. Welche Freude | »Tgei plascher ch'ins haveva!«

the primary school in Danis. And for the same number of years as church organist, he was awarded the Benemerenti medal by the Pope. Altogether, he crossed the bridge many, many thousands of times.

On a frosty day in January 1913, Salvator Caduff was carried over Robert Maillart's bridge to the font. As a child, he used the bridge and witnessed its downfall, awed by the swift construction of its successor. And throughout his life, he showed a keen interest in the ups and downs of life in his village that remains untouched by the sun's rays from early November to mid-February. Just a few months before his death, wishing to pass on what he knew, Salvator wrote his memoirs: »Entginas notizias da Danis-Tavanasa« | Some notes on Danis-Tavanasa. His daughter Mengia let me read them, these notizias by the teacher and organist Salvator Caduff-Zortea who for some years now has been dark earth in the cemetery, over in Danis.

»Mo purs havevan nus fetg paucs a Tavanasa« | Few livelihoods in Tavanasa were based purely on farming, he writes. There was too little usable land between the steep slopes and the Rhine. But: »Nus lessen menziunar che bunamein tuts havevan sper lur mistregn, tgi in pli grond e tgi in pli pign menaschi puril, ina sanadeivla economia.« | We'd like to mention that almost everyone had a modest piece of farm land in addition to their craft trades, a healthy economy. The lack of grazing and arable land, and thus the need to make a living beyond agriculture, was at the root of Tavanasa's industriousness, shaping its diverse and, as Salvator found, healthy economy with craft businesses that served mostly rural customers from across the region. Ignaz Derungs, following in Caduff's footsteps as chronicler of Danis-Tavanasa (www.ils-derungs.ch), has put together a list of no less than 31 trades practiced in Danis-Tavanasa in the past, some of which have survived into the present.

»Lein entscheiver entadem Tavanasa« | Let's begin at the furthermost end in Tavanasa, suggests Salvator Caduff. »Leu steva sper la val dil Lochli la casa dalla fam. Alig.« | At the mouth of the Val dil Lochli stood the house of the Alig family. The Aligs were newcomers. In the mid-19th century, a Josef Anton Alig from Obersaxen came down to the shady village of Tavanasa, settled in the Lochli, learned Romansh, changed his name to Giusep Antoni and began to cast bransinas, cowbells, in copper and tin. His son, also Giusep Antoni, married Turtè Riedi from Danis and carried on his father's business, and two of their sons are also learning the craft of bell foundry. Ethnic prejudice seems to have been alien to the Aligs, who were happy to occasionally work with two Yenish. Johann Friedrich Moser moved from Obervaz to Schnaus and later to Danis. The bells were marked Moser Danis, but often also Moser Alig or Alig Moser. His son Ferdinand also worked with the Aligs, although as a travelling founder he was often away working at his customer's homes. »NOTICE The undersigned announces that this year he will once again repair defective cowbells, make and sell new ones large and small, as well as accepting scrap metal or household items in lieu of payment. – Anyone sending work should give their exact address and state precisely what's to be done. – On 12 November, the undersigned will come in person to the St Martin's Market in Disentis with a range of cowbells that will be on sale for very reasonable prices. – The cowbells are widely known and bear the inscription A l i g. Gius. Ant. Alig, cowbell founder, Tavana-

Inserat in der Gasetta Romontscha, 16.10.1924

Advert in the Gasetta Romontscha, 16.10.1924

Avis.

Il suttascret fa enconuschent, ch' el culi era uonn

Brensinas

ruttas, fetschi novas, vendi grondas e pintgas, bratti talas encunter metall veder, sco era rauba de casa.

Tgi che termetta lavurs vegli indicar exact sia adressa e dir co ins giavischa la lavur.

La **fiera S. Martin a Mustér,** dils 13 de Nov., vegn il suttascret sez cun ina partida brensinas, ch' ein d' haver per fetg raschuneivels prezis.

Las brensinas ein lunsch entuorn enconuschentas e portan la firma Alig. 2 2

Gius. Ant. Alig, cula-brensinas, **Tavanasa.**

Auch in die Familiengeschichte von Lehrer Salvator ist der Gewerbefleiss von Tavanasa eingeschrieben. »La fatschenta dils cuntschapials ei vegnida fundada da Giachen Fidel Caduff.« | Der Gerbereibetrieb ist von Giachen Fidel Caduff (1829-1900) gegründet worden, mit Hilfe eines deutschen Gerbers namens Sigfrid, hält er in den notizias fest, und: »Ils cuntschapials da Tavanasa fagevan treis sorts curom.« | Die Tavanaser Gerber stellten drei Sorten Leder her. Der Schulmeister zählt sie auf: Sohl-, Ober- und Riemenleder. Er kennt noch alle Arbeitsgänge aus eigener Anschauung, die verschiedenen Phasen der Grubengerbung, die Herstellung der Lohbrühe mit Tannen- und Eichenrinde, das Zerkleinern der Rinde in der Stampfmühle draußen an der Val dil Tschart. Drei Generationen Caduff haben die Gerberei mitten im Dorf betrieben, 1954 ist damit Schluss, die industrielle Konkurrenz mit ihren schnelleren chemischen Verfahren ist übermächtig. Der letzte Gerber der Surselva, Salvators Vater, stirbt 1968 mit 93 Jahren. Aber ein wenig bleibt der Sohn der Lederbranche treu. Die Lehrerlöhne sind bescheiden, zusammen mit seiner Frau führt er zuhause ein kleines Schuhgeschäft.

Der Sommer 1927 bringt viel Regen, Unwetter und Hagelschlag, das ganze Schweizer Mittelland wird arg in Mitleidenschaft gezogen. Auch in den Bündner Bergen ist der Boden von den anhaltenden Regenfällen durchtränkt und schwer.
Am 22. September kommt ein starker Föhnwind auf, das Barometer fällt und fällt. Am Samstagabend, 24. September stauen sich schwere, schwarz-rote Wolken, ein warmer, heftiger Regen, dracca da favugn setzt ein, ergießt sich pausenlos die ganze lange Herbstnacht vorab über die rechte Talseite der oberen Surselva. Rinnsale sind unversehens reissende Wildbäche, der Rhein wälzt sich tosend als braune, stetig steigende Flut durchs Tal. Der Sonntagmorgen dämmert unter Blitz und Donner und der Himmel entleert sich weiter, in draccar e tunar diluvial senza fin e misericordia, ohne Ende und Erbarmen.
Zur Mittagszeit wagt sich der Stiel- und Rechenmacher Barclamiu Zortea im strömenden Regen auf die Brücke, unter sich die reißende, braune Flut des hochgehenden Flusses, als plötzlich der rechte Talhang krachend und donnernd birst, die Mure kommt und die kleine Welt von Tavanasa innert Augenblicken eine andere ist. »La part occidentala dil vitg ei totalmein devastada.« | Der westliche Dorfteil ist völlig zerstört, schreibt Salvator Caduff, drei Wohnhäuser, ein Ökonomiegebäude, zwei Heuställe, die schöne, 1905 erbaute Brücke nach Danis, die Kantonsstraße von Tavanasa nach Trun und die Linie der Rhätischen Bahn. Aber das Schrecklichste ist, dass dieses Ereignis sieben Einwohner von Tavanasa das Leben gekostet hat. Rechenmacher Zortea ist einer von ihnen.
Turtè Parpan-Beer, Barrierenwärterin und Mutter von sechs Kindern, ereilt der Tod im Wärterhäuschen. Eben hat sie nach der Zugsdurchfahrt die Schranke wieder hochgekurbelt. Der knapp verschonte Zug wird vor Darvella steckenbleiben. Jodocus und Nicolaus Caduff, 16- und 15-jährig, müssen sterben, weil sie ihren Eltern und den sechs Geschwistern zu Hilfe eilen wollen. Diese aber sind dem Unheil entkommen, nur Sekunden bevor das Haus von der Mure erdrückt und weggetragen wird. Ils Naglers heißt die Familie im Dorf, weil der Großvater hier eine Nagelschmiede betrieb. Vater Clau Caduff ist nach dem Verlust der beiden Söhne ein gebrochener Mann, stirbt zwei Jahre nach dem Unglück.

Der Rechenmacher Giusep Spescha (Foto: Giorgio von Arb)

Rake-maker Giusep Spescha (Photo: Giorgio von Arb)

sa.« (notice in the Gasetta Romontscha, 16 October 1924, right picture, page 57) In Salvator's notes I read that when they were not casting bells, father Alig and his sons would be welding buckets and tubs out of tin – and that they were enthusiastic anglers, the first people in Tavanasa to use not just nets and traps but also rod and line. In Dautras, where the hydroelectric power station was built in the 1960s, Toni Alig, a slight man, is said to have caught a lake trout weighing 18 pounds. More than anything else, however, Salvator cherished the memory of the bright sounds he heard when, as a boy, he would pass the Lochli when the bransiners were testing a set of freshly cast bells. What a joy | »tgei plascher ch'ins haveva!«

The industriousness of Tavanasa is also written into Salvator's own family history. »La fatschenta dils cuntschapials ei vegnida fundada da Giachen Fidel Caduff.« | The tannery was founded by Giachen Fidel Caduff (1829-1900), with the help of a German tanner by the name of Sigfrid, he recorded in his notizias: »Ils cuntschapials da Tavanasa fagevan treis sorts curom« | The Tavanasa tanners made three kinds of leather, which he lists as leather for soles, uppers and straps. He knew all of the work processes from watching them himself, the various phases of pit tanning, the preparation of the tanning fluid with pine and oak bark, the breaking down of the bark at the stamping mill in the Val dil Tschart. Three generations of Caduffs ran the tannery in the middle of the village until 1954 when the business was overpowered by competition from the faster chemical processes used in industrial tanning. The last tanner in the Surselva, Salvator's father, died in 1968 aged 93. But his son did remain true to the leather trade: a teacher's wages being modest, he ran a small shoe shop at home with his wife.

The summer of 1927 brought rain, storms and hail, with all of central Switzerland badly affected. In the mountains of Grisons, too, the ground was saturated by the endless rain. On 22 September, a strong foehn wind blew and the barometer kept falling. On the evening of Saturday, 24 September, heavy dark red clouds piled up, a warm heavy rain started to fall, dracca da favugn, pouring down ceaselessly throughout the night onto the right-hand side of the upper Surselva valley. Rivulets suddenly became raging torrents, the Rhine roared down the valley as an ever-rising brown flood. Sunday dawned with thunder and lighting and the downpour continued, in draccar e tunar diluvial senza fin e misericordia, without end and without mercy. At midday, the handle- and rake-maker Barclamiu Zortea had ventured out on to the bridge in the pouring rain, below him the brown torrent of the swelling river, when suddenly, with a crashing and a thundering, the right-hand side of the valley burst and down came the landslide, transforming the little world of Tavanasa in an instant. »La part occidentala dil vitg ei totalmein devastada.« | The western part of the village was totally destroyed, writes Salvator Caduff, three houses, a farm building, two hay barns, the beautiful bridge to Danis built in 1905, the cantonal highway from Tavanasa to Trun, and the tracks of the Rhaetian Railway. Most terribly of all, however, this event cost seven inhabitants of Tavanasa their lives. Rake-maker Zortea was among them.

Turtè Parpan-Beer, level-crossing attendant and mother of six children, was killed in her attendant's shelter, having just cranked the barrier back up. The train, having narrowly made it through, got stuck outside Darvella. Jodocus and Nicolaus Caduff, aged 16 and 15, died while going to help their parents and six siblings

Früher Holzsteg, bis 1906 Early wooden bridge, until 1906

Maillart-Brücke, gebaut 1905/1906 Maillart's bridge, built 1905/1906

Nach der Rüfe, 1927 After the landslide, 1927

Der Murgang vernichtet auch das Wohnhaus und die Glockengießerei der Aligs im Lochli innert wenigen Augenblicken. Giusep Antoni, seine Frau Turtè und Sohn Luis, Maurer und Vater von fünf Kindern, kommen um. Eine Tochter kann sich mit knapper Not aus dem Schlamm retten, von dem sie schon erfasst worden war. Auch Sohn und Glockengießer Toni überlebt - Kuhglocken der Marke Alig werden aber künftig keine mehr entstehen. Wo das Haus stand, erinnert fortan ein Kreuz mit Emailschild an die sieben Toten von 1927: »Rughei per lur olmas!« | Betet für ihre Seelen! Vor ein paar Jahren wurde das Lochli-Kreuz ans andere Rheinufer versetzt, neben eine Orientierungstafel und die letzten Trümmer der Maillart-Brücke. Gutgemeinte Gedankenlosigkeit hat aus dem Memento ein Exponat gemacht (Bild rechts).

Am 25. September 1927 birst der rechte Talhang krachend und donnernd, die Mure zerstört einige Gebäude und die 1906 gebaute Maillart-Brücke. Sieben Menschen kommen bei dem Unglück ums Leben

Niemand war schon immer hier – wie die Aligs war auch Rechenmacher Barclamiu Zortea ein Zugezogener, ein Immigrant. »El era vegnius alla entschatta dil 20avel tschentaner ord in vitg cul num Zortea el Tirol« | Er war anfangs des zwanzigsten Jahrhunderts aus einem Dorf namens Zortea im Südtirol gekommen. Der Moment ist günstig, die Auftragslage für einen Produzenten von Rechen und Werkzeugstielen könnte besser nicht sein. Denn die Rhätische Bahn baut von 1910-12 die Strecke von Ilanz nach Disentis, Hundertschaften italienischer Arbeiter pickeln und schaufeln im Tal. Zorteas Werkzeugstiele sind für ihre Qualität bekannt und verkaufen sich fast von selbst. Nach Barclamius Tod in der Rüfe führt dessen einziger Sohn Giusep den Betrieb weiter. Er stirbt 1940, sein Lehrling und Nachfolger Giusep Friberg verlegt die Werkstatt für ein paar Jahre nach Danis hinüber. »El ha era giu in emprendist« | Auch er hatte einen Lehrling, il Giusep Spescha. Der stellt seine Werkzeugstiele und Heurechen wie einst die Zorteas wieder in Tavanasa her, am untern Dorfende Richtung Chur, und fortan verkündet in der »Gasetta Romontscha« eine Kleinanzeige mit schöner Regelmäßigkeit: »Monis da tuttas sorts furnescha Giusep Spescha, Tavanasa.« | Stiele aller Art liefert Giusep Spescha, Tavanasa.

Am 15. März 2000 stirbt der letzte Stiel- und Rechenmacher von Tavanasa. Schon zehn Jahre zuvor hatte er zum Schriftsteller Vic Hendry gesagt: »A Tavanasa ei miu igniv, vi Danis mon jeu sin santeri.« | In Tavanasa ist mein Daheim, nach Danis rüber gehe ich auf den Friedhof.

Reste der zerstörten Maillart-Brücke *Remains of destroyed Maillart bridge*

Im Lochli, 2017 *Im Lochli, 2017*

Das Kreuz zum Gedenken der sieben bei der Rüfe Umgekommenen stand ursprünglich im Lochli. Vor einigen Jahren wurde es auf die andere Rheinseite transloziert

The cross commemorating the seven people killed by the landslide originally stood in the Lochli. Several years ago, it was moved to the other bank of the Rhine

– who had already escaped, just seconds before the house was crushed by the landslide and carried away. The family were known in the village as ils Naglers because the grandfather had been a nailsmith. Having lost his two sons, Clau Caduff was a broken man, dying two years later.

In an instant, the landslide also destroyed the house and the bell foundry of the Aligs in the Lochli. Giusep Antoni, his wife Turtè and his son Luis, a bricklayer and father of five children, were killed. A daughter was able to make a narrow escape from the mud that had already engulfed her. The son and bell founder Toni also survived – but there were to be no more cowbells marked ALIG. Where the house once stood, a cross with an enamel sign commemorated the seven dead of 1927: »Rughei per lur olmas!« | Pray for their souls! A few years ago, the Lochli cross was moved to the other bank of the Rhine, standing beside an information panel and the last vestiges of Maillart's bridge. Well-intentioned thoughtlessness has turned the memorial into an exhibit (Picture above right).

No one has always been here – like the Aligs, rake-maker Barclamiu Zortea was also a newcomer, an immigrant. »El era vegnius alla entschatta dil 20avel tschentaner ord in vitg cul num Zortea el Tirol.« | He came in the early 20th century from a village called Zortea in South Tyrol. The moment was opportune, the situation for a producer of rakes and handles for all manner of tools could not have been better: between 1910 and 1912, the Rhaetian Railway built the line from Ilanz to Disentis, with hundreds of Italian workers picking and shovelling in the valley. Zortea's tool handles were known for their quality and sold like hot cakes. After Barclamiu's death in the landslide, his only son Giusep carried on the business. He died in 1940, his apprentice and successor Giusep Friberg moved the workshop over to Danis for a few years. »El ha era giu in emprendist« | he too had an apprentice, il Giusep Spescha. Like the Zorteas before him, he made his tool handles and hay rakes in Tavanasa, at the bottom end of the village towards Chur, and a regular notice in the Gasetta Romontscha proclaimed: »Monis da tuttas sorts furnescha Giusep Spescha, Tavanasa.« | Handles of all kinds, supplied by Giusep Spescha, Tavanasa.

On 25 September 1927, the right-hand slope of the valley thunderously collapsed. The landslide destroyed several buildings and the bridge built by Maillart in 1906. Seven people lost their lives

Tavanasa's last handle- and rake-maker died on 15 March 2000. Ten years earlier, he had already told the writer Vic Hendry: »A Tavanasa ei miu igniv, vi Danis mon jeu sin santeri« | Tavanasa is my home, I go over to Danis to the cemetery.

Scursanida · Abkürzung | Robert Kruker: Verkehr auf alten Wegen

Bei Dardin-Mulin beginnt der alte Weg nach Baditschus, unterhalb von Breil/Brigels. Hier wurde 2017 die wegbegleitende Trockenmauer erneuert (rechts im Bild)

The old path to Baditschus, below Breil/Brigels, begins at Dardin-Mulin. In 2017, the dry wall along the path was restored (on the right in the picture)

»Jeder Bauer ist gezwungen, im Winter trotz Sturm und Schneegestöber auf schmalem Fußsteg oder auf gänzlich schneeverwehten, oft noch steilem Pfade täglich zweimal seine warme Dorfbehausung zu verlassen: am Morgen in der Frühe und abends nach vier Uhr, um den relativ oft weiten Weg (eine Viertel- bis eine Stunde) zu seinem Viehstall anzutreten, die Tiere zu füttern und den Milchertrag in seinen hölzernen Tansen nach Haus zu tragen«, schrieb um 1916 Pater Karl Hager, der sich auf seinen wissenschaftlichen und fotografischen Streifzügen auch verschiedentlich in der Gegend von Danis-Tavanasa und Dardin aufhielt.

Ils trutgs, die alten Pfade zu den aclas, den Gadenstätten, und zwischen den Dörfern wurden bis um 1960 rege genutzt. Die Fußwege verbanden Danis-Tavanasa nicht nur mit den Fraktionen von Dardin und mit Breil/Brigels, sie führten auch zu den Höfen südlich des Rheins, von denen zwei, nämlich Vali und Cathomen zur Gemeinde Breil/Brigels gehörten. Axenstein, Bellaua, Platta und Hanschenhaus lagen zwar auf Obersaxer Gebiet, wurden aber bis 1959 vom Postboten aus Tavanasa bedient, das für diese Weiler damals der wichtigste Versorgungspunkt war.

Im Sommer 1914 war die Bergstraße bis Breil/Brigels fertiggestellt und ersetzte die bisherige Hauptverbindung über Waltensburg. 1928, nach dem Bau der neuen Rheinbrücke, löste das Postauto die Kutsche ab. Motorfahrzeuge, die in Graubünden erst nach 1925 zugelassen wurden, waren noch äußerst selten. Im Winter konnten sie aufgrund der schwierigen Straßenverhältnisse meist nicht fahren. Dann verkehrten weiterhin geschlossene Zweispännerschlitten. In besonderen Situationen wie zum Beispiel einer Geburt in Danis, zu der die Hebamme aus Breil/Brigels gerufen wurde, fuhr der Pferdeschlitten eines privaten Fuhrhalters.

Über die Nutzung der alten Wege gibt es viele überlieferte Geschichten. Auf dem felsdurchsetzten trutg dalla Val Cuschina (1) von Dardin nach Danis-Tavanasa verschwand einst ein kleines Mädchen. Es wurde nicht mehr gefunden, und das Gerücht von einem bösen Tobelgeist machte die Runde. Bistgaun Cathomen (Jahrgang 1951) erinnert sich, dass er und seine gleichaltrige Schwester Marcella als Kinder immer zu zweit von Dardin hinunter nach Tavanasa zum Coiffeur oder zum Schuhmacher gegangen seien. Den trutg dalla Val Cuschina (1) hätten sie gemieden und den sanfteren trutg da Tschitschè (2) nach Danis genommen, den heutigen Wan-

Scursanida · Shortcut | Robert Kruker: Travelling the old paths

»In winter, in spite of storms and blizzards, every farmer is obliged to leave his cosy village dwelling twice a day, on narrow tracks or on snowbound and often steep paths: once early in the morning and once in the evening after four o'clock, walking between fifteen minutes and an hour to his cattle shed to feed the animals and then carry the milk home in wooden tubs.« This account was written around 1916 by Pater Karl Hager, a scholar and photographer who spent various periods in Danis-Tavanasa and Dardin.

Until around 1960, ils trutgs, the old paths to the aclas, the pastures, and between the villages, were much used. As well as linking Danis-Tavanasa with Dardin and Breil/Brigels, they also led to the farms south of the Rhine, of which two (Vali und Cathomen) belonged to the municipality of Breil/Brigels. Although the hamlets of Axenstein, Bellaua, Platta and Hanschenhaus belonged to the municipality of Obersaxen, until 1959 they were served by a postman from Tavanasa, at the time the main supply point for these communities. In the summer of 1914, the mountain road was completed as far as Breil/Brigels, replacing the previous route via Waltensburg. In 1928, after the construction of the Rhine bridge, the horse-drawn post carriage was replaced by a car. Motor vehicles, not authorized in Grisons until after 1925, were still extremely rare. In winter, when difficult road conditions often ruled out their use, closed horse-drawn sledges continued to operate. In special situations, for example if the midwife from Breil/Brigels was called to a birth in Danis, a privately owned horse-drawn sledge was used.

Many stories have survived about the old pathways. Once, a small girl disappeared on the rocky trutg dalla Val Cuschina (1) from Dardin to Danis-Tavanasa. She was never found and there were rumours of an evil spirit in the ravine. Bistgaun Cathomen (born 1951) remembers how, as children, he and his sister Marcella would go down from Dardin to Tavanasa together to the hairdresser or the cobbler. They would avoid the trutg dalla Val Cuschina (1) and take the gentler trutg da Tschitschè (2) to Danis, today's hiking trail. Ursulina Cavegn-Livers from Dardin remembered always walking the same way down via Tschitschè to cook lunch for the guests at

Robert Kruker in Dardin-Mulin bei der Trockenmauer-Baustelle (September 2017)

Robert Kruker at the dry wall construction site in Dardin-Mulin (September 2017)

derweg. Ursulina Cavegn-Livers aus Dardin berichtete, dass sie jeweils denselben Weg über Tschitschè hinunterlief, um in der Ustria Staziun in Tavanasa das Mittagessen für die Gäste zu kochen. Bistgaun Fidel Cathomen von Dardin war von 1924-34 Lehrer und Organist in Danis. Nicht selten sollen ihn am Frühstückstisch erst die Kirchenglocken an die sonntägliche Pflicht erinnert haben. Er sei dann auf dem trutg dalla Val Cuschina (1) hinuntergeeilt, um fünf Minuten später seinen Platz auf der Orgelbank einzunehmen.

Ustria Staziun in Tavanasa. Between 1924 and 1934, Bistgaun Fidel Cathomen from Dardin was a teacher and organist in Danis. Often, sitting at the breakfast table, the ringing of the church bell would remind him of his Sunday duties. Then he would hurry down the trutg dalla Val Cuschina (1), taking his seat on the organist's bench five minutes later.

West of the Val Cuschina, the trutg da Chistatscha (3) leads to the place where the young men would fling the Schibettas, burning wooden discs, into the night sky

Scursanida, die alte via sut Baselgia von Danis zur Ustria Sur Punt um 1900

Scursanida, the old via sut Baselgia from Danis to Ustria Sur Punt, circa 1900

Westlich der Val Cuschina führt der trutg da Chistatscha (3) zur Stelle, wo die Jungmannschaft jeweils am ersten Samstag der Fastenzeit die Schibettas, die glühenden Scheiben, in den Nachthimmel über Danis schlagen. Der Weg heißt weiter oben trutg da Canariel (4). Das war einst ein Prozessionsweg bis hinauf nach Baditschus bei Breil/Brigels. Ihn nutzten früher auch die in Breil/Brigels stationierten Soldaten auf ihrem Weg in den Urlaub. Singend sollen sie in Einerkolonne zu Tal marschiert sein. Das letzte Wegstück vor der Rheinbrücke, die via sut Baselgia (5) oder la scursanida, die Abkürzung genannt, ist heute nicht mehr durchgehend begehbar. Dieser Teil zwischen der Casa Parvenda und der Ustria Sur Punt ist dem Bau von Häusern und der Umfahrung von Tavanasa gewichen.

Das dichte Netz von Verbindungs- und Flurwegen ist Vergangenheit. Meliorationsstraßen mit Hartbelag führen zu den verbliebenen Ställen, viele der alten Pfade sind eingewachsen oder wurden gar ausplaniert. Einige der alten Verbindungswege sind jetzt Wanderwege, Ortskundige nutzen einige Seitenpfade als Abkürzung oder beim Pilzsammeln.

Der alte Weg von Dardin-Mulin nach Baditschus unterhalb von Breil/Brigels, laut Inventar der historischen Verkehrswege »von regionaler Bedeutung mit Substanz«, wird in den Jahren 2017-2018 restauriert. Unter sachkundiger Leitung und unterstützt von Zivilschutz und Dorfvereinen repariert die Werkgruppe der Gemeinde die wegbegleitenden Trockenmauern und legt alte Pflästerungen wieder frei (Bild auf Seite 62).

over Danis on the first Saturday in Lent. Further up, the path is called trutg da Canariel (4). In earlier times, this was a procession route up to Baditschus near Breil/Brigels. This same path was also formerly used by the soldiers stationed in Breil/Brigels on their way home. They are said to have marched in single file, singing, down into the valley. The last stretch of path before the Rhine bridge, known as via sut Baselgia (5) or la scursanida (the shortcut), no longer exists in its entirety, as the section between the Casa Parvenda and Ustria Sur Punt has been used to build houses and the Tavanasa ring-road.

The dense network of connecting and cross-country paths is a thing of the past. Today, surfaced roads lead to the remaining cattle sheds and many of the old paths have become overgrown or even been bulldozed. Some of the old connecting routes are now hiking trails, some locals use certain secondary paths as shortcuts or when gathering mushrooms.

The old path from Dardin-Mulin to Baditschus beneath Breil/Brigels, listed in the index of historical transport routes as being »of substantial regional significance«, is being restored in the years 2017-18. Under expert guidance and supported by the Zivilschutz (civil defence authority) and village associations, the municipal working group is repairing the dry walls along the path and uncovering old cobblestones. (Picture on page 62).

La punt | Arno Camenisch

Ich war mir nie sicher, ob die Brücke auch halten würde, seit man uns Kindern vom Steinschlag 1927 erzählte, der die alte Brücke mitgerissen hatte. Im Rhein steht heute noch ein Stein wie ein Haus, der vom Steinschlag vor bald hundert Jahren zeugt. Wir standen mitten auf der Brücke und schauten runter auf den Stein und wieder rauf zum Ochlifelsen über dem Dorf, wo sich der Steinschlag gelöst hatte. Später behauptete jemand aus der Klasse, in der Brücke sei Sprengstoff, und im Fall der Fälle würde die Brücke gesprengt werden, und wieder machten wir Kinder große Augen und stellten uns vor, beim Kiosk neben der Zapfsäule zu stehen, mit den Tornistern auf dem Rücken, um in den Abgrund zu schauen. Und auch vor wenigen Jahren war ich mir nicht sicher, ob die Brücke noch da wäre, wenn ich das nächste Mal ins Dorf fahren würde, weil die Brücke hätte abgerissen werden sollen. Und als ich kürzlich über die Brücke spaziert bin, um meinen Vater zu besuchen, der auf der anderen Seite der Brücke wohnt, dachte ich an das Zitat von Mark Twain, der einmal gesagt haben soll: »Ich bin ein alter Mann und habe viele Kümmernisse erlebt – von denen die meisten nie geschehen sind« und war froh, dass die Befürchtung über das Verschwinden der Brücke, die seit klein auf da war, nie eingetroffen ist.

Im Roman »Ustrinkata« beschreibt Arno Camenisch den Tag, an dem die Ustria Helvezia vor einigen Jahren zum letzten Mal geöffnet war

I was never sure whether the bridge would hold up, ever since we were told as children about the landslide of 1927 that took the old bridge with it. Even today, a rock the size of a house stands in the Rhine, testimony to that event of almost a century ago. We would stand in the middle of the bridge and look down at the rock and then up at the Ochlifelsen above the village where the landslide had started. Later, someone in my class claimed that the bridge was fitted with explosives and that if the worst came to the worst, it would be blown up, and us children were wide-eyed again, imagining ourselves standing by the kiosk next to the petrol pump, satchels on our backs, looking down into the abyss. And even a few years ago, I wasn't sure whether the bridge would still be there the next time I came to the village, as it was scheduled for demolition. And when I walked over the bridge on a recent visit my father who lives on the other side I thought of the quote from Mark Twain, who is supposed to have said: »I am an old man and have known a great many troubles, but most of them never happened.« And I was glad that my worries about the bridge disappearing, that have been there since childhood, never came true.

In his novel »Ustrinkata« Arno Camenisch describes the day several years ago when Ustria Helvezia opened for the last time

Jeu erel mai segirs, sche la punt tegni propi, dapi ch'ei havevan raquintau a nus affons dalla bova dil 1927 che haveva destruiu la punt veglia. El Rein stat oz aunc in crap sc'ina casa che dat perdetga dalla bova d'avon beingleiti 100 onns. Nus stevan a mesa punt e miravan giu sil crap e puspei si sil grep digl Ochli, nua che la bova era rutta ora. Pli tard haveva enzatgi ord mia classa pretendiu ch'ei hagi dinamit ella punt, e che la punt vegni el cass dils cass siglientada ell'aria, e puspei fagevan nus egls gronds e s'imaginavan da star spel kiosc cul tancadi e mirar giuaden el precipezi. Ed era avon paucs onns erel jeu buca segirs, sche la punt seigi aunc leu la proxima gada ch'jeu seigi el vitg, perquei ch'ei levan scarpar giu ella. E cu jeu sun ius dacuort sur la punt vi per visitar miu bab, che stat da l'autra vart dalla punt, hai jeu patertgau vid in citat da Mark Twain, che havevi detg inaga »Ich bin ein alter Mann und habe viele Kümmernisse erlebt – von denen die meisten nie geschehen sind« e sun staus leds, ch'il quitau, che la punt, ch'era da pign ensi leu, savessi esser svanida, ei mai daventaus realitad.

El roman »Ustrinkata« descriva Arno Camenisch il davos di che l'Ustria Helvezia ei stada aviarta

Alice Cathomas

*1921 in Tavanasa. War Kindergärtnerin und letzte Wirtin der Ustria sur Punt. Lebt jetzt im Altersheim in Obersaxen.

Born 1921 in Tavanasa. Former nursery school teacher and last landlady of Ustria sur Punt. Now lives in an old people's home in Obersaxen.

Die Brücke würde fehlen! Stellt euch vor, man würde in Mailand den Dom abreißen ...

The bridge would be missed. Imagine what it would be like if they demolished the cathedral in Milan ...

La punt muncass - e lu aunc co! Mettei avon ch'il dom da Milaun vegness scarpaus giu ...

Samuele Gianoli

*1975 in Lugano. Baut als Schreiner alte Häuser um. 2016 Kauf der Ustria Sur Punt. Seither mit dem Umbau beschäftigt.
Born 1975 in Lugano. A carpenter who converts old houses. Bought Ustria sur Punt in 2016 and has been working on it ever since.

Ich habe einen Narren gefressen an der Ustria sur Punt, die lange ungenutzt leerstand. Die Ustria gab es lange bevor Versell die schöne, neue Brücke baute. Ich möchte sie erhalten und wieder mit Leben füllen.
I'm crazy about the Ustria sur Punt, which long stood empty and unused. The inn existed long before Versell built his beautiful new bridge. I'd like to preserve it and fill it with life again.

Flurin Caduff

*1979, aufgewachsen in Danis-Tavanasa. Freischaffender Opernsänger, wohnt in Luzern.
Born in 1979, grew up in Danis-Tavanasa. Opera singer, lives in Lucerne.

Die Brücke bedeutet für mich Heimat, Wurzeln, Erinnerungen und Freundschaften.
For me, the bridge means home, roots, memories and friendships.
La punt munta per mei ragischs e patria, regurdientschas ed amicezias.

URBAN
ASIAN
DIVISION

Robert Hugi

*1953 in Solothurn. Holztransporte in Tavanasa. Seit 1987 Schäfer.
Born 1953 in Solothurn. Timber haulier in Tavanasa. Since 1987, he has been a shepherd.

Wenn die Brücke nicht mehr wäre – wo sollte ich dann meine Schafe rübertreiben?
If the bridge were no longer there – where would I herd my sheep across?
Sche la punt fuss buca pli – nua duess jeu lu ir suravi cun mias nuorsas?

David Flepp

*1975 in Danis. Lebt in Dardin und arbeitet als Schauspieler, Kabarettist und Sprachförderer.

Born 1975 in Danis. Lives in Dardin and works as an actor, cabaret artist and language coach.

Während sieben Jahren habe ich tagtäglich zweimal die Brücke überquert.
Morgens in Richtung Tavanasa Station, abends in Richtung Tumvi, meinem Zuhause.
For seven years, I crossed the bridge twice a day:
in the morning towards Tavanasa Station, in the evening towards Tumvi, where I lived.
Duront siat onns sun jeu ius di per di duas ga sur la punt.
La damaun per ir vi Tavanasa-staziun, la sera per turnar a casa vi Tumvi.

Armin Caduff

* 1949 in Tavanasa. Komponist und Sänger. Maestro des Solistenensembles La Compagnia Rossini mit weltweiten Auftritten. Lebt in Danis.

Born 1949. Composer and singer. Director of the vocal ensemble La Compagnia Rossini with concerts worldwide. Lives in Danis.

»Ich höre den Rhein« ist der Titel einer meiner Kompositionen. Darin hört man unterschiedliche Tempi und variantenreiche Klänge, wie ich den Rhein auf der Brücke in meiner Kindheit erlebte.

»Ich höre den Rhein« (I hear the Rhine) is the title of one of my compositions. It features different tempi and various sounds, as I experienced the Rhine on the bridge in my childhood.

»Jeu audel il Rein« ei il tetel dad ina da mias cumposiziuns. En quella aud'ins differents tempos e varionts accords sco quei ch'jeu udevel il Rein sin la punt da miu temps d'affonza.

Bruno Cathomas

*1965 in Laax. War als Kind immer in den Sommerferien in Tavanasa bei Alice Cathomas, seiner Tante.
Schauspieler. Lebt in Köln und Berlin.
Born 1965 in Laax. Spent his childhood summer holidays in Tavanasa with his aunt Alice Cathomas.
Actor. Lives in Cologne and Berlin.

Meine Perspektive der Brücke in Tavanasa ist eine untersichtige.
Da ich als Kind oft die Ferien in der Ustria sur Punt verbringen konnte, wohnte ich quasi unter der Brücke. Mir war es unter Strafe verboten worden den Brückenbogen zu besteigen. Aber oben am Bogen war eine kleine Höhle sichtbar und die musste erkundet werden. Strafe hin oder her; der Bogen wurde unter großer Angst bestiegen. In der Höhle befand sich tatsächlich etwas: eine halbverweste Katze. Tausendmal schneller als ich hochgeklettert war, und voll von mystischer Hexenangst, war ich wieder unten. Ich habe diesen Brückenbogen nie mehr betreten.

My view of the bridge in Tavanasa is from below.
As a child, I was lucky enough to spend many holidays at Ustria sur Punt, living more or less under the bridge. Climbing the arch was a punishable offense. But I could see a little cavity up there, and I had to explore it, whatever the consequence. With great trepidation, I climbed the arch. And there really was something in the cavity: a half decayed cat. I came down a thousand times faster than I'd gone up, full of a mystical fear of witches. I never set foot on the bridge arch again.

Mia vesta dalla punt ei ina da sutsi.
Da buob vai jeu astgau far savens vacanzas en l'Ustria sur Punt. Quei vul dir, ch'jeu habitavel lu quasi sut la punt. Da reiver sigl artg dalla punt vevani scumandau a mi strentgamein. Aber si leu vevi ina caverna ch'ins veseva da sutsi - e quella ha stuiu vegnir sondada, castitg vi ni neu. Da reiver si leu hai jeu temiu spitachel. Ella caverna hai jeu lu viu, tgei ch'ei era: in gat miez smarschiu. Melli ga pli spert che da reiver ensi e cun la sgarschur da ver ina stria el dies, sun jeu staus engiu. Quei ei stau l'emprema e la davosa ga ch'jeu sun revius sigl artg dalla punt.

Versell-Brücke | Versell's bridge

Unsere Brücke | Rita Cathomas-Bearth: Der Weg zu Nossa Punt

Sieben Brücken führen auf dem Gebiet von Tavanasa über den jungen Rhein. Die Zufahrt zur heutigen Feriendestination Breil/Brigels war fast 90 Jahre lang allein über die enge einspurige Versell-Brücke möglich. Als die breiter und schwerer werdenden Trucks die Fußgänger immer prekärer an den Rand drängten und das alte Bauwerk an seine Grenzen kam, handelte der Kanton. Er baute eine breitere Brücke und bot die ausgediente der Gemeinde an. Gemeindebehörde und Stimmbürger aber lehnten entschieden ab. Die koste zu viel und könne abgerissen werden.

Weder ein Gutachten, das der Brücke hohe Schutzwürdigkeit attestierte, weder Heimatschutz noch Denkmalpflege, noch die Leute von Tavanasa konnten dieses Nein verhindern. Für den Kanton war das ein Dilemma. Ihm fehlte die gesetzliche Grundlage, um eine ausgediente Brücke zu behalten. Gleichzeitig hätte ein Abbruch gegen das eigene Heimatschutzgesetz verstoßen, das von Privaten und selbstverständlich auch vom Kanton den Erhalt von wertvollen Bauten fordert.

Die Versell-Brücke ist formschön und stark. Sie überstand Hochwasser und erinnert an den tragischen Murgang von 1927. Nun wurde sie zu unserer Brücke – Nossa Punt. Zum Wahrzeichen nicht nur für Tavanasa

In dieser verfahrenen Situation ließ ich mich, als ehemalige Präsidentin des Bündner Heimatschutzes, von der heutigen Geschäftsführerin überzeugen, mich für den Erhalt der Versell-Brücke einzusetzen. Eine Woche nach der Volksabstimmung war der Verein gegründet. Drei Ortsansässige, ein Enkel von Walter Versell und ich als Präsidentin bildeten den Vorstand. Ein Unterstützungskomitee mit prominenten Persönlichkeiten aus ganz Graubünden stärkte uns den Rücken. Mit Flyern, Mundpropaganda, einem Internetauftritt und Medienberichten leisteten wir Überzeugungsarbeit. »Nossa Punt«

Our Bridge | Rita Cathomas-Bearth: The path to Nossa Punt

In the Tavanasa area, seven bridges lead over the upper reaches of the Rhine. For almost 90 years, today's tourist destination of Breil/Brigels could only be accessed via the narrow, one-lane bridge built by Walter Versell. As the trucks became wider and heavier, leaving less and less space for pedestrians and pushing the old structure to its limits, the canton took action. It built a wider bridge and offered the old one to the municipality. But the municipal authorities and electors were decidedly against: the bridge would cost too much and could be demolished.

Versell's bridge is elegant and strong. It has resisted floods and recalls the tragic landslide of 1927. Now it has become our bridge – Nossa Punt. An emblematic landmark, not only for Tavanasa

Neither a report confirming the bridge's protection status, nor the Offices of Cultural Heritage and Historic Monuments, nor the people of Tavanasa could prevent the negative vote. This presented the canton with a dilemma. It lacked a legal basis on which to keep a disused bridge. At the same time, its demolition would have violated the canton's own Law on the Protection of Nature and Cultural Heritage that calls on both private individuals and the canton itself, of course, to preserve valuable structures.

In this deadlocked situation, the directors of the Grisons Cultural Heritage Office persuaded me, as its former president, to work to save Versell's bridge. A week after the referendum, the association was founded, its managing committee consisting of three local residents, a grandson of Versell's, and myself as president. A committee of prominent figures from all over the Canton of Grisons supported us. Using flyers, word-of-mouth, a website and media coverage, we were able to persuade

NOSSA PUNT

wurde zum Signet. Innert kürzester Zeit traten rund zweihundert Mitglieder, Gönnerinnen und Gönner dem Verein bei. Viele erkannten, dass die alte Brücke mehr ist als eine Ruine.
Bei Bund und Kanton stießen wir auf offene Ohren. Die zuständigen Ämter waren froh, dass unser Verein bereit war, mit einer Stiftung das Bauwerk zu übernehmen. Die Instandsetzung wurde finanziell gesichert. Fußgängerinnen und Fußgänger zwischen Tavanasa und Danis atmeten auf, denn so wurde ihnen ein unangenehmer Umweg über die stark befahrene neue Brücke erspart.
Die Versell-Brücke ist formschön und stark. Die ganze Entwicklung von Breil zum Feriendorf hat sie mitgetragen. Tausende von Tonnen Baumaterialien, Feriengäste, Soldaten des Flablagers, Einheimische, die kamen und gingen. Sie überstand Hochwasser und erinnert an den tragischen Murgang, der sieben Menschenleben forderte und die Maillart-Brücke zerstörte. So wurde die Versell-Brücke zu unserer Brücke – Nossa Punt. Zum geschichtsträchtigen Denkmal. Zum Wahrzeichen nicht nur für Tavanasa.
Die Sicherung der Versell-Brücke für weitere Jahrzehnte ist das Ergebnis einer fruchtbaren Zusammenarbeiter all jener Kräfte, die davon überzeugt sind, dass neben dem Neuen, das unsere Zeit prägt, auch alte erhaltenswürdige Werke ihre Bedeutung haben und unsere Gesellschaft bereichern.

Ich danke allen, die mitgewirkt haben, dass die Versell-Brücke heute noch steht, dass sie instandgesetzt wurde, dass dieses Buch entstehen konnte und allen, die sich weiterhin um dieses Bauwerk kümmern.

people. »Nossa Punt« became a brand. Within a short period, around 200 members and patrons joined the association. Many people realized that the old bridge is more than just a ruin.
The federal and canton authorities were interested in what we had to say. The responsible offices were glad that our association was prepared to take on the bridge through a foundation. Funding for the restoration was secured. People needing to walk from Tavanasa to Danis breathed a sigh of relief, as this saved them an unpleasant detour via the busy road over the new bridge with its traffic.

Versell's bridge is strong and elegant. It was instrumental in the development of Breil into a holiday village. It has seen the passage of thousands of tons of building materials, holidaymakers, soldiers from the air defence barracks, locals coming and going. It survived floods and recalls the tragic landslide that cost seven lives and destroyed its predecessor. In this way, Versell's bridge became our bridge – Nossa Punt. It became a historical monument, a landmark not just for Tavanasa.

The preservation of Versell's bridge for future generations is the result of a constructive joining of forces in the belief that alongside the new that shapes our times, old works also possess relevance and enrich our society.

I would like to thank all those who helped to ensure that Versell's bridge is still standing today, everyone involved in the restoration work and in producing this book, and all those who will continue to look after the bridge in the future.

Sep Cathomas

*1945 in Breil/Brigels. War Mitinhaber eines Architekturbüros, Mitglied des Gemeindevorstands und Gemeindepräsident sowie Präsident des Gemeindeverbandes Surselva. Ehemaliges Mitglied des Schweizer Nationalrats. Lebt in Breil/Brigels. Präsident der Fundaziun Nossa Punt.
Born 1945 in Breil/Brigels. Formerly partner in a firm of architects, member of the municipal council and mayor, president of the Surselva regional council. Former member of the Swiss National Council. Lives in Breil/Brigels. President of Fundaziun Nossa Punt.

Die Versell-Brücke bildet für mich seit meiner Jugend die Verbindung zwischen meinem Geburtsort Brigels und der »großen Welt«.
For me, Versell's bridge has always been the connection between Brigels, where I was born, and the »big, wide world«.
Per mei fuorma la punt da Versell dapi mia giuventetgna la colligiaziun denter miu vitg nativ Breil ed il »vast mund«.

Marc Cathomen

*1969 in Danis. Arbeitet als IT-Unternehmer. Lebt in Danis. Vorstandsmitglied der Uniun Nossa Punt.
Born 1969 in Danis. Works in IT. Lives in Danis. Board member of Uniun Nossa Punt.

Diese Brücke ist gut für die Autos. Die Versell-Brücke hingegen ist und bleibt das Bindeglied zwischen Danis und Tavanasa.
This bridge is good for cars. But Versell's bridge is and remains the link between Danis and Tavanasa.
Nossa Punt ei la coligiaziun denter las duas parts dad in vitg Danis-Tavanasa.

Christian Alig

*1969 in Chur, aufgewachsen in Landquart. Wohnt in Schluein (bei Ilanz). Leiter der Region Surselva bei der Graubündner Kantonalbank, Ilanz. Stiftungsrat und Aktuar der Fundaziun Nossa Punt.
Born 1969 in Chur, grew up in Landquart, lives in Schluein (near Ilanz). Regional director for Surselva at Graubündner Kantonalbank, Ilanz. Trustee and actuary of Fundaziun Nossa Punt.

Die Versell-Brücke wirkt trotz ihres Alters modern und sie passt sich wunderbar in die Landschaft ein. Deshalb setze ich mich für den Erhalt der Brücke ein.
In spite of its age, Versell's bridge looks modern and it fits wonderfully into the landscape. That's why I'm working to preserve the bridge.
La punt da Versell ei moderna, malgrad ses onns, e s'adattescha excellent alla cuntrada. Perquei s'engaschel jeu era per mantener ella.

Rita Cathomas-Bearth

* 1949 in Rabius. Übersetzerin und Projektleiterin. Lebt in Chur. Präsidentin der Uniun und Stiftungsrätin der Fundaziun Nossa Punt.

Born in 1949 in Rabius. Translator and project manager. Lives in Chur. President of Union Nossa Punt and trustee of Fundaziun Nossa Punt.

Die neue Conzett-Brücke ist eindrücklich und notwendig. Dass die wertvolle Versell-Brücke erhalten werden konnte, freut mich aber ebenso.

Jürg Conzett's new bridge is impressive and necessary. But I am also very glad it was possible to save Versell's precious bridge.

La nova punt da Conzett ei impressiunonta e necessaria. Jeu hai era grond plascher che la biala punt da Versell ha saviu vegnir mantenida.

Alex Liechti

*1955 in Kitale, Kenia. Wirtschaftswissenschaftler (lic. oec. publ.). Lebt in Effretikon. Vorstandsmitglied der Uniun Nossa Punt.

Born 1955 in Kitale, Kenya. Economist (lic. oec. publ.). Lives in Effretikon. Board member of Uniun Nossa Punt.

Mein Großvater, Walter Versell, hat mir viel über seine Brücken erzählt. Die Schilderungen vom Hochwasser, welches das Lehrgerüst der soeben betonierten Tavanaser Brücke wegzuschwemmen drohte, faszinierte mich besonders. Ich freue mich sehr über alle Anstrengungen, die zu ihrer Erhaltung geführt haben.

My grandfather, Walter Versell, told me many things about his bridges. His stories of the flood that threatened to wash away the formwork for the freshly poured concrete of the bridge in Tavanasa was especially fascinating. I'm very glad about all the efforts that have led to its preservation.

Marlis Caduff

*1953 in Tavanasa. Hat bei der Gemeindeverwaltung Ilanz/Glion gearbeitet. Lebt in Tavanasa. Aktuarin der Uniun Nossa punt.

Born 1953 in Tavanasa. Worked in the municipal administration of Ilanz/Glion. Lives in Tavanasa. Actuary of the Nossa punt association.

Ich hab's mal nachgerechnet: Allein während der Schulzeit bin ich wohl 7000 mal über die Brücke gelaufen. Das bleibt im Gedächtnis. Das prägt einen Menschen. Diese Brücke ist für mich die »Golden Gate« von Tavanasa!

I once worked out that during my school years alone, I crossed the bridge 7000 times. That's not something you forget. It makes an impression. For me, this bridge is the »Golden Gate« of Tavanasa!

Jeu hai inaga quintau suenter: Sulet mo duront il temps da scola sundel jeu ida varga 7000 ga a pei sur la punt vi. Quei rest'en memoria. Quei dat tempra al carstgaun. Quella punt ei per mei la »Golden Gate« da Tavanasa.

Rita und Daniel Carigiet

Rita *1974 in Untervaz. Daniel *1971 in Dardin. Landwirte. Leben und arbeiten gemeinsam in Tavanasa. Daniel ist Vizepräsident von Uniun und Stiftung Nossa Punt.
Rita born 1974 in Untervaz. Daniel born 1971 in Dardin. Farmers. Live and work together in Tavanasa. Daniel is vice president of both Uniun and Fundaziun Nossa Punt.

Wir wollten diese Brücke unbedingt behalten. Nun ist sie zu einem wunderschönen und sehenswerten Fußweg geworden. Wir sind glücklich und stolz auf »unsere Brücke«.
We wanted to keep this bridge at all costs. Now it has become an attractive pedestrian route. We are happy and proud of »our bridge«.
Nus levan salvar per tut prezi quella punt. Ussa eis ella daventada ina bellezia e remarcabla senda. Nus essan ventireivels e loschs da »nossa punt«.

Flurina Carigiet

*2004 in Tavanasa. Geht in Danis zur Schule.
Born 2004 in Tavanasa. Goes to school in Danis.

Die Brücke gehört zu Tavanasa. Uns gefällt, dass jetzt keine Autos mehr darüber fahren.
The bridge belongs to Tavanasa. We like the fact that cars no longer use it.
La punt s'auda tier Tavanasa. A nus plai che negins autos van pli sur la punt vi.

Aurelio Carigiet

*2005 in Tavanasa. Geht in Danis zur Schule.
Born 2005 in Tavanasa. Goes to school in Danis.

Die Brücke war schon immer einzigartig. Durch die Beleuchtung bekam sie jetzt noch den letzten Schliff.
The bridge has always been unique. The lights have added a finishing touch.
La punt ei adina stada unica. Cun l'illuminaziun ha ella survegniu la davosa glimada.

Jonas Beer

*1998 in Danis. 2011-17 Gymnasium in Disentis. 2017 Maturaarbeit über die Versell-Brücke. Lebt in Danis.
Born 1998 in Danis. 2011-17: high school in Disentis. 2017: final-year project about Versell's bridge. Lives in Danis.

Als Schüler bin ich oft genug über die Versell-Brücke gesprintet, um den Zug nach Disentis, also zur Schule nur ja nicht zu verpassen. Das bleibt im Gedächtnis.
Over the years, I often had to sprint over Versell's bridge so as not to miss the train to school in Disentis. That's something you don't forget.
Sco scolar sun jeu currius savens avunda sur la punt da Versell vi per puder tier il tren viers Mustér e per gie buca munchentar la scola. Quei resta en memoria.

Charly Caduff

*1954 in Tavanasa. Elektriker. Lebt in Danis.
Born 1954 in Tavanasa. Electrician. Lives in Danis.

Die Brücke ist für mich so wichtig wie der Eiffelturm in Paris. Mit der Brücke sind schöne Erinnerungen an meine Jugend verbunden. 2016 habe ich mit einer Lichtaktion das Augenmerk auf die Brücke gelenkt.
For me, this bridge is as important as the Eiffel Tower in Paris. I associate it with good memories from my youth. In 2016, I drew attention to the bridge with a light installation.
La punt ei per mei schi impurtonta sco la tuor d'Eiffel a Paris. Ella regorda mei a bials muments da mia giuventetgna. 2016 hai jeu fatg in'acziun d'illuminaziun per mussar, con biala che la punt ei.

Ciril Friberg

*1943, aufgewachsen in Tavanasa im Haus De Latour. Ehemaliger Lehrer. Präsident des Forum cultural Breil. Lebt in Danis.
Born 1943, grew up in Tavanasa at Haus De Latour. Former teacher. President of Forum cultural Breil. Lives in Danis.

Mit der Rettung der Versellbrücke werden wir zurückversetzt in die 1950er-Jahre. Da konnten die Fußgänger, der Geißhirt und der Schafhirt mit ihren Herden die Brücke in aller Ruhe, ohne Verkehr traversieren. Jetzt können alle den Leichtverkehr wieder genießen.
The restoration of Versell's bridge takes us back to the situation in the 1950s. Back then, goatherds and shepherds could cross the river with their herds in peace, without traffic. Now cyclists and pedestrians can enjoy that again.
Cul salvament dalla punt da Versell turnein nus anavos els onns 1950. Cheu savevan ils pedunzs, il nurser e cavrer cun lur muntaneras, traversar la punt senza disturbi dad autos. Uss san tuts guder il traffic lev e la ruasseivladad.

Franzisca Beer-Schmed

*1929 in Breil/Brigels. Lebt seit Jahrzehnten in Tavanasa (neben der Brücke).
Born 1929 in Breil/Brigels. Lives for decades in Tavanasa (next to the bridge).

Tavanasa wäre ohne die Brücke nicht mehr Tavanasa. Es wäre dort ein Loch, wenn die Brücke weg käme.
Without the bridge, Tavanasa wouldn't be Tavanasa. If the bridge were taken away, it would leave a hole.
Senza la punt fuss Tavanasa buca Tavanasa pli. Sche la punt vegness naven, restass mo aunc ina ruosna anavos.

Venanzi Caduff

*1946 in Tavanasa. War LKW-Fahrer. Wohnt in Tavanasa.
Born 1946 in Tavanasa. Retired lorry driver. Lives in Tavanasa.

Als die Kraftwerke rheinaufwärts gebaut wurden, war es für uns Kinder ein Spektakel, wenn die Sattelschlepper mit Baumaschinen wegen geringer Durchfahrtshöhe vor der Brücke Halt machen mussten. Da standen wir stundenlang und haben zugeschaut, wenn hier alles ab- und drüben wieder aufgeladen wurde.
When the power stations were being built, it was a spectacle for us children when articulated trucks bringing heavy machinery had to stop below the bridge because it was too low for them to pass. We stood and watched for hours as everything was unloaded and then loaded back onto the trucks on the other side.
Cu che l'ovra electrica ei vegnida baghegiada dadens Tavanasa, stuevan ils camiuns cun las maschinas da baghegiar tener dado il vitg avon la punt-fier, perquei ch'il passadi era memia bass. Nus buobs stevan lu uras ô e miravan e miravan, co ei scargavan giu las maschinas avon la punt e cargavan si ellas puspei da l'autra vart.

Rezia und Jasmin Derungs

Rezia *1971 in Tavanasa. Seit 2006 als Briefträgerin in Danis-Tavanasa. Jasmin *2003 in Tavanasa. Geht in Danis zur Schule.
Rezia born 1971 in Tavanasa. Delivering the mail in Danis-Tavanasa since 2006. Jasmin born 2003 in Tavanasa. Goes to school in Danis.

Die Brücke verbindet Tavanasa mit Danis, so werden zwei Dörfer zu einem.
The bridge connects Tavanasa with Danis, turning two villages into one.
La punt colligia Tavanasa cun Danis, aschia fa ella ord dus vitgs in.

Maria Caduff-Cahenzli

*1923 in Pilavarda, Obersaxen. Kam als Magd nach Tavanasa und lebt dort seit Jahrzehnten.
Born in Pilavarda, Obersaxen. Came to Tavanasa as a maidservant and has been living there for decades.

Ich kann es wirklich nicht mehr zählen, wie oft ich über diese Brücke gegangen bin. Es wäre doch jammerschade, wenn man sie abgerissen hätte.
I really can't put a number on the amount of times I've crossed this bridge. It would have been such a shame if it had been demolished.
Jeu sai propi buc, contas gadas ch'jeu sun ida tochen oz sur quella punt vi. Ei fuss bein stau donn e puccau, sch'ei vessen scarpau giu ella.

Mengia Frei-Caduff

*1952 in Tavanasa geboren und dort aufgewachsen. Lehrerin. Wohnt in Schlieren (und – im früheren Elternhaus – in Tavanasa).
Born 1952 in Tavanasa, where she grew up. Teacher. Lives in Schlieren (and, in her old family home, in Tavanasa).

Erinnerungen an meinen Schulweg nach Danis und an Treffen mit meinen Freundinnen auf dem Sportplatz »Isletta« (unterhalb der Brücke) kommen auf.
It brings back memories of walking to school in Danis and meeting my friends on the »Isletta« playing field (below the bridge).
Regurdientschas da mia via da scola e da sentupadas cun mias amitgas giu l' isletta revivan.

Rosita und Gieri Friberg

Rosita *1946 in Tavanasa (bei der Brücke). Telefonistin, Bürokauffrau. Mitinitiantin für Nossa Punt. Gieri *1949 in Cavardiras. Gieri hat als Plattenleger gearbeitet. Rosita und Gieri wohnen in Tavanasa (bei der Brücke).
Rosita, born 1946 in Tavanasa (near the bridge). Switchboard operator, office clerk. Co-initiator of Nossa Punt. Gieri, born 1949 in Cavardiras. Retired paver. Rosita and Gieri live in Tavanasa (near the bridge).

Wenn die Brücke wegkommt, dann gehen wir auch! (Rosita und Gieri, 2015)
If the bridge goes, then we go too! (Rosita & Gieri, 2015)
Sche la punt vegn naven mein nus era. (Rosita e Gieri, 2015)

Erhalt der Versell-Brücke | Simon Berger: Suche nach neuer Trägerschaft

Ebenso unverhofft wie das Unwetter am 25. September 1927 zum Neubau der Vorderrheinbrücke führte, so wurde dem neuen Denkmalpfleger des Kantons ab September 2014 die Aufgabe zuteil, für diese Brücke eine neue Trägerschaft zu finden um sie vor dem Abriss zu bewahren. Die Rahmenbedingungen dafür waren gegeben: Für das vorherrschende Verkehrsaufkommen war die fast ein Jahrhundert alte Brücke nicht mehr geeignet. Das Tiefbauamt des Kantons hatte daher beantragt, flussaufwärts eine neue Brücke zu bauen. Die Regierung bewilligte den Neubau. Außerdem sollte die Schutzwürdigkeit der alten Brücke abgeklärt und eine Trägerschaft für die Brücke und deren Unterhalt gefunden werden. Die Suche sollte die Denkmalpflege übernehmen. Der (Termin-)Druck war groß, denn die alte Brücke hätte abgerissen werden sollen, wenn bis zur Fertigstellung der neuen Brücke kein Träger für Erhalt und Unterhalt der alten gefunden worden wäre.

Ein Expertengutachten bestätigte die Schutzwürdigkeit der Brücke. Technisch, historisch, aber auch für das Ortsbild ist die Brücke herausragend. Die Gemeinde Breil/Brigels lehnte eine Übernahme der alten Brücke ab. Also musste eine private Trägerschaft gefunden werden. Dank der Initiative von vielen engagierten Personen wurde der Verein Nossa Punt gegründet. Daraus ging die Stiftung Nossa Punt hervor, welche die Brücke und den Unterhalt der Brücke übernahm. Mit Subventionsgeldern von Bund und Kanton wurde die Brücke 2017 instandgesetzt und unter Schutz gestellt. Als Fußgänger- und Fahrradverbindung steht sie nun zukünftigen Generationen zur Verfügung.

Tavanasa, alte und neue Brücke, Juli 2016 (links)

Tavanasa, old and new bridge, July 2016 (left)

Saving Versell's bridge | Simon Berger: Looking for a new owner-operator model

Just as unexpectedly as the storm of 25 September 1927 led to the construction of a new bridge over the Rhine, in September 2014 the Canton's new officer for historic monuments was given the task of finding a new owner-operator model for this bridge in order to save it from demolition. After nearly a century, it was no longer able to deal with the traffic load. The Canton's civil engineering department was thus commissioned to build a new bridge upstream and the plans were then authorized by government. In addition, the Office of Historic Monuments was tasked with clarifying the protection status of the old bridge and finding a new owner-operator model – under serious time pressure, as the old bridge was scheduled for demolition if no new arrangement had been found by the time the new one was completed. An expert assessment confirmed the bridge's preservation status. Technically and historically, but also within the local landscape, the bridge is outstanding. The municipality of Breil/Brigels refused to take on the old bridge, triggering the search for a private owner-operator. Thanks to the initiative of many committed individuals, the association »Nossa Punt« came into being, leading in turn to the creation of the Nossa Punt Foundation as the new owner responsible for operating and maintaining Versell's bridge. With subsidies from government at federal and canton level, the bridge was restored in 2017 and given listed status, securing it as a connection for pedestrians and cyclists for generations to come.

Maillart · Versell · Menn | Jürg Conzett: Die Versell-Brücke und ihre Vorgängerin

Die Maillart-Brücke von 1905

In Tavanasa baute Robert Maillart (1872-1940) als Ingenieur und Unternehmer seine erste Dreigelenk-Hohlkastenbrücke mit ausgeschnittenen Bogenzwickeln (Bild Seite 98). 1905 erstellt, war sie der Prototyp seiner berühmten Dreigelenk-Bogenbrücken, etwa der Salginatobelbrücke bei Schiers (Seite 97, linkes Bild), der Rossgrabenbrücke bei Hinterfultigen (BE) oder der Arvebrücke bei Vessy (GE). Ausgehend von der Vorderrheinbrücke in Tavanasa schuf Maillart einen neuartigen, ebenso wirtschaftlichen wie eleganten Brückentyp, dessen Form er aus den besonderen Eigenschaften des »armierten Betons« herleitete (Bild auf Seite 98).

Was die Brücke in Tavanasa derart bemerkenswert machte, war das Ausschneiden der Bogenzwickel. Hohlkastenbrücken mit geschlossenen Seitenwänden hatte Maillart schon vorher gebaut, erstmals bei der Innbrücke in Zuoz. Hier hatte Maillart eine dünne Bogenplatte auf ein leichtes und kostengünstiges Lehrgerüst gegossen. Diese »Schale« wurde anschließend durch Längswände versteift, die ihrerseits die Fahrbahn trugen. Das Ganze bildete einen dünnwandigen Hohlkasten, der mit den steinernen Vorbildern der frühen Betonbogenbrücken nichts mehr gemein hatte. Vorteilhaft war insbesondere, dass das Gewicht der Längswände und der Fahrbahnplatte während des Betonierens von der bereits erhärteten Betonplatte des Bogens aufgenommen wurde, sodass das Gerüst nur für

Die Vorderrheinbrücke Tavanasa ist ein Hauptwerk des bedeutenden Ingenieurs Versell, sie steht an einem brückengeschichtlich bedeutenden Ort, zeigt einen ungewöhnlich schlanken Betonbogen und läßt sich in der Entwicklung des Betonbrückenbaus präzis zwischen Maillart und Menn platzieren

Maillart · Versell · Menn | Jürg Conzett: The Versell's bridge and its predecessors

Maillart's bridge of 1905

In Tavanasa, the civil engineer Robert Maillart (1872-1940) built his first three-hinged box-girder bridge with open spandrels (picture on page 98). Completed in 1905, it was the prototype for his famous three-hinged arch bridges, including the Salginatobel Bridge in Schiers (left picture on page 97), the Rossgraben Bridge in Hinterfultigen, and the Arve Bridge in Vessy. Beginning with this bridge over the Rhine in Tavanasa, Maillart created a new type of bridge, both economical and elegant, its form derived from the special properties of reinforced concrete (picture on page 98).

What made the bridge in Tavanasa so remarkable was the open spandrels. Maillart had already built box-girder bridges with closed spandrels, beginning with the bridge over the Inn in Zuoz. Here, he cast a thin arch panel on a light, inexpensive centring.

This »shell« was then stiffened by longitudinal walls that in turn carried the road deck. Overall, this formed a thin-walled box girder that no longer had anything in common with the stone bridges on which the early concrete arch bridges were modelled.

One particular advantage was that while they were being poured, the weight of the longitudinal walls and the road deck were born by the already hardened concrete panel of the arch so that the scaffolding could be calculated for the load of the arch

The bridge over the upper reaches of the Rhine at Tavanasa is a chef-d'oeuvre of the eminent civil engineer Walter Versell, standing at a historically significant location. With its unusually slender concrete arch, it occupies a position in the development of concrete bridge building between Maillart and Menn

Robert Maillart, Salginatobelbrücke, 1930

Robert Maillart, Salginatobel Bridge, 1930

Walter Versell, Brücke über den Vorderrhein in Tavanasa, 1928
Besichtigung mit Jürg Conzett (links) und Sep Cathomas, Juli 2016

Walter Versell, bridge over the Upper Rhine in Tavanasa, 1928
Site visit with Jürg Conzett (left) and Sep Cathomas, July 2016

Christian Menn, Ponte Nanin, Brücke über den Fluss Moesa, 1967

Christian Menn, Ponte Nanin, bridge over the River Moesa, 1967

Maillart-Brücke, Tavanasa, 1906 (oben). Foto: Carl Lang, Fotostiftung Graubünden. Ganterbrücke, Christian Menn, 1980 (rechts)

Maillart's bridge, Tavanasa, 1906 (top). Photo: Carl Lang, Fotostiftung Graubünden. Ganter Bridge, Christian Menn, 1980 (right)

die Last des Bogens allein bemessen werden musste. Dieses Prinzip bewährte sich derart gut, dass Maillart kurz darauf eine ähnliche Konstruktion über die Thur bei Billwil (TG) ausführen konnte. Von außen sahen diese Brücken noch vergleichsweise konventionell aus; das statisch komplexe Innenleben der Hohlkastenkonstruktion blieb verborgen.

Nach einiger Zeit zeigten sich in den Längswänden der Innbrücke Zuoz nahe der Auflager kleine und an sich harmlose Risse, die Maillart veranlassten, weiter über die Funktionsweise seiner Erfindung nachzudenken. Er beschloss, in der Vorderrheinbrücke Tavanasa die zu Rissen neigende Wandpartie einfach wegzulassen. So gelang es Maillart, einen Mangel seiner früheren Brücken zu verbessern und dabei obendrein noch Material zu sparen.

Architektonisch wurde die Tavanasa-Brücke zur ersten Ikone des »Maillart-Bogens«, der die Gelenke durch Zuspitzen des Betons zum Ausdruck bringt und dafür in den Viertelspunkten der Spannweite eine große Stärke aufweist.

So schuf Maillart in Tavanasa eine Brücke, die die weitere Entwicklung des Betonbrückenbaus im 20. Jahrhundert wesentlich mitprägte. Ihr Konstruktionsprinzip ist in abgewandelter Form noch in der Ganterbrücke am Simplonpass (Christian Menn, 1981) zu erkennen (Bild Seite 99).

Zerstörung der Maillart-Brücke 1927 und das Projekt der neuen Brücke

Gegen Ende September 1927 regnete es in der Ostschweiz anhaltend. Am 25. September erreichte der Wasserstand des Rheins eine Höchstmarke. Gleichzeitig löste sich überraschend aus der unscheinbaren »Val da Lochli« ein heftiger Murgang, der

alone. This principle proved so successful that Maillart was able to use a similar construction for his bridge over the Thur at Billwil soon after. From outside, these bridges still looked comparatively conventional; the load-bearing complexities of the box-girder construction remained hidden.

After a while, small and essentially harmless cracks appeared in the longitudinal walls close to the abutments of the bridge in Zuoz, prompting Maillart to think more about the way his invention worked. For the bridge over the Rhine in Tavanasa, he decided to simply leave out the part of the wall that tended to crack. In this way, Maillart managed to improve a defect of his earlier bridges, while also saving material.

In architectural terms, the Tavanasa Bridge became the first icon of the »Maillart arch«, an arch which shows its greatest structural strength in the quarter-points of the span and tapers from there towards the joints, giving them a strong visual presence.

In this way, Maillart created a bridge in Tavanasa that helped to shape the development of concrete bridge building in the 20th century. Its construction principles can still be identified in altered form in the Ganter Bridge on the Simplon Pass (Christian Menn, 1981, right picture).

Destruction of Maillart's bridge in 1927 and the project for a replacement

Towards the end of September 1927, there was a period of sustained rain in eastern Switzerland. On 25 September, the level of the Rhine reached a new high. At the same time, a heavy landslide occurred in the Val da Lochli that destroyed three

drei Häuser und Ställe zerstörte, Straße und Bahn verschüttete, schließlich Maillarts Brücke seitlich erfasste und sie auf das linke Rheinufer warf. Sieben Menschen wurden durch diesen Murgang getötet.

Noch im Oktober organisierte Kantonsingenieur Johann Solca ein Wettbewerbsverfahren für den Neubau der Brücke und formulierte die »Leitsätze für die Ausführung der Brücke als Betonbogenbrücke«. Die neue Brücke wurde flussaufwärts der Mündung der bedrohlichen Val da Lochli platziert, sie überquerte den Vorderrhein und die Rhätische Bahn in Hochlage.

Wesentliche konstruktive Einzelheiten legte Solca bereits in den Leitsätzen fest: So sollten die Kämpferpfeiler und Widerlager in Beton mit Steinverkleidung ausgeführt werden und die Brüstungen aus Beton bestehen; diese waren alle 2,5 bis 3 Meter mit runden Öffnungen zu versehen. Bis zum 25. November 1927 hatten Ingenieure und Unternehmer Zeit, Projekte mit Pauschalofferten auszuarbeiten.

Am Wettbewerb beteiligten sich drei kantonale und drei außerkantonale Unternehmungen, darunter auch der Ingenieur Robert Maillart mit der Bauunternehmung Prader & Cie (Zürich und Chur). Sowohl Maillart als auch Versell schlugen Bogenbrücken mit 50,80 Metern Spannweite vor. Maillarts Wettbewerbsentwurf war ein versteifter Stabbogen mit zwei Mal fünf Feldern über dem Bogen, sehr ähnlich seiner Valtschielbrücke von 1925. Walter Versell, der sich mit der Unternehmung B. & C. Caprez aus Chur und Arosa am Wettbewerb beteiligte, schlug dagegen einen schmalen schlanken eingespannten Bogen mit zwei Mal drei Feldern über dem Bogen vor (Bild Seite 102). Nach einigen Diskussionen mit der Gemeinde, die zunächst eine niedrige Brücke favorisierte, für die Maillart auch einen Entwurf lieferte, entschied sich das kantonale Tiefbauamt am 5. Februar schließlich für Versells Entwurf und der Kleine Rat Graubündens bestätigte diesen Entscheid am 2. März 1928. Maßgebend für diesen Entscheid dürften Kostengründe gewesen sein. In der Tat ist Versells Brücke äußerst materialsparend. Die Stärke des 3 Meter breiten Bogens variiert zwischen 42 Zentimeter im Scheitel und 73 Zentimeter am Kämpfer. Nehmen wir Maillarts versteiften Bogen der Valtschielbrücke als Maßstab für seinen Tavanasa-Entwurf, wäre sein Bogen zwar nur etwa halb so stark wie Versells Bogen gewesen, wäre aber mit 4,40 Meter genauso breit geworden wie der trogförmige Versteifungsträger der Fahrbahn. Wegen der gegenüber der Versell-Brücke geringeren Knicksteifigkeit des Bogens benötigte Maillart zwei Mal vier Pfeilerscheiben zwischen Bogen und Fahrbahn, während Versell sich mit zwei Mal zwei schmaleren Scheiben begnügen konnte. Versells Brücke war dadurch hinsichtlich Materialverbrauch Maillarts Brücke ebenbürtig, Versells Lehrgerüst des Bogens war schmaler, und Versells Brücke wies wegen der geringeren Anzahl Pfeilerscheiben weniger Schalfläche als Maillarts Entwurf auf. So erhielten Versell und Caprez den Zuschlag für die neue Brücke.

Der Brückenbau dieser Jahre zeichnete sich durch ein hochentwickeltes Gefühl für Sparsamkeit aus, was den Entwürfen von Maillart, Versell und ihren Zeitgenossen einen prägnanten Charakter verleiht. Versell schrieb denn auch an Ingenieur Mirko Roš, den Direktor der eidgenössischen Materialprüfanstalt, der seine Brücke mit einer Belastungsprobe prüfte, es sei für ihn nicht darum gegangen mit den sehr schlanken Dimensionen technische Rekorde zu brechen, sondern vielmehr nach niederen Baukosten zu streben.

houses and barns, buried the street and the railway tracks, and then struck Maillart's bridge sideways on, throwing it onto the left-hand bank of the Rhine. Seven people died.

In October, the canton's engineer-in-chief Johann Solca organized a competition for a new bridge and formulated »Guidelines for Execution of the Bridge as a Concrete Arch Bridge«.

The new bridge was to be located upstream of where the hazardous Val da Lochli entered the valley, spanning the Rhine and the Rhaetian Railway at an elevated level. Key constructive details were laid down in these guidelines: the abutments and abutment pillars were to be realized in stone-clad concrete and the balustrades would be concrete, perforated by round openings every 2.5 or 3 metres. Civil engineers and firms had until 25 November 1927 to develop costed tenders. Three cantonal offices and three non-cantonal firms took part, including Robert Maillart with the firm of Prader & Cie (Zürich and Chur).

Both Maillart and Versell suggested arched bridges with spans of 50.80 metres. Maillart's entry was a stiffened deck arch bridge with two times five spandrels over the arch, very similar to his Valtschiel Bridge of 1925. Walter Versell, who entered the competition with the firm of B. & C. Caprez (Chur and Arosa), proposed a narrow, slender fixed arch with two times three spandrels over the arch (picture on page 102). After discussions with the local authorities, who initially favoured a low bridge, for which Maillart also submitted a design, on 5 February the cantonal civil engineering office chose Versell's design and the government of Grisons conformed this decision on 2 March 1928.

This decision is likely to have been based largely on costs, since Versell's bridge is very economical in its use of materials. The thickness of the 3 metres wide arch varies between 42 centimetres at the apex and 73 centimetres at the abutments. Compared with Versell's Tavanasa design, the stiffened arch of Maillart's Valtschiel Bridge was only half as thick as Versell's, while measuring 4.40 metres across, the same width as the stiffening beam of the road deck.

Because of the arch's lower buckling resistance compared with Versell's bridge, Maillart needed two times four spandrel columns between the arch and the road deck, whereas Versell could make do with two times two thinner columns. In terms of material usage, then, Versell's bridge was thus a match for Maillart's; Versell's formwork for the arch was narrower and because it had less spandrel columns the overall surface of its formwork was also smaller. For these reasons, Versell and Caprez were chosen to build the new bridge.

At this time, bridge building was marked by a highly developed sense of economy, lending designs by Maillart, Versell and their contemporaries a distinctive character. As Versell wrote to Mirko Roš, the director of the Swiss Material Testing Laboratory who carried out a load test on the bridge, his main aim was not to break records with a very narrow bridge, but to minimize construction costs.

Especially interesting is the way Versell determined the arch's form and dimensions. The arch is a hinge-less or fixed arch. The distribution of the bending moments in a fixed arch depends on the varying section heights. With no aids, one must begin by making geometrical assumptions, which are then checked, corrected, checked again, etc. – a laborious procedure.

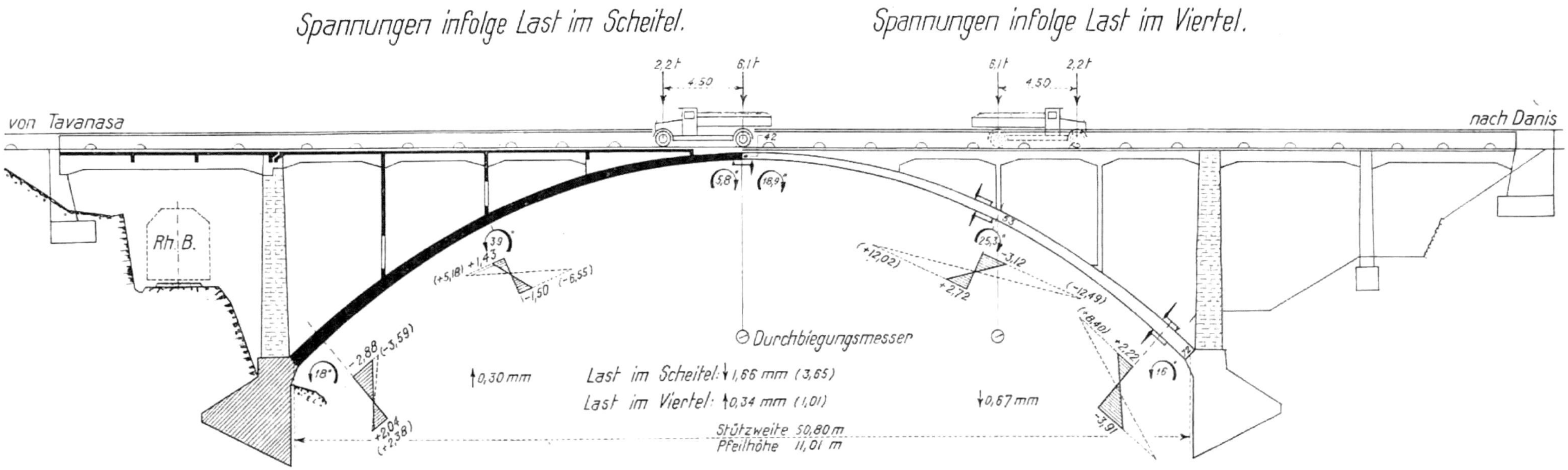

Versell-Brücke mit Ergebnissen der Belastungsversuche, 1928

Versell's bridge with results of load testing, 1928

Interessant ist vor allem, wie Versell Form und Dimension des Bogens bestimmte. Der Bogen ist ein sogenannter »eingespannter Bogen«. Die Verteilung der Biegemomente in einem eingespannten Bogen hängt vom Verlauf der Querschnittshöhen ab. Ohne Hilfsmittel muss man zuerst geometrische Annahmen treffen, diese überprüfen, korrigieren, neu prüfen usw. – eine arbeitsintensive Fleißarbeit. Um diese zu verkürzen, benutzte Versell ein Tabellenwerk von A. Strassner zur Formfindung eingespannter Bogen. Auf dieser Näherung aufbauend, berechnete er die wirkliche Drucklinie, korrigierte diese aus ästhetischen Gründen um einige Zentimeter und führte dann die definitive Berechnung nach der Elastizitätstheorie durch.

In den Zeiten der ersten Betonbogenbrücken berechnete man eingespannte Bögen so, als ob sie die ganze Last der Brücke allein aufnehmen würden. Die Fahrbahn hatte nur die »lokale« Funktion, die Lasten zu den Stützwänden und damit in den Bogen zu leiten. Bald wurden jedoch die Grenzen dieser vereinfachten Annahme spürbar, denn wenn sich ein schlanker Bogen verformt, überträgt sich diese Verfor-

To speed up this procedure, Versell used a set of tables for calculating the form of fixed arches published by A. Strassner. On the basis of this approximation, he calculated the actual line of thrust, corrected it by a few centimetres for aesthetic reasons, and then carried out the definitive calculation according to the rules of elasticity theory.

In the days of the first concrete arch bridges, fixed arches were calculated as if they would bear the entire load of the bridge on their own. The road deck only had the »local« function of transferring the loads to the supporting walls and thus into the arch.

The limitations of this simplified assumption soon became clear, however, because when a slender arch becomes deformed, this deformation is also transferred to the road deck, calling forth a resistance in it that decreases the load on the arch but that also places added stresses on the road deck. Versell calculated that the bending moments were divided between arch and road deck in a ratio of ¾ to ¼.

Versell-Brücke, kurz nach der Fertigstellung. Foto: W. Goetz, Chur

Versell's bridge, soon after completion. Photo: W. Goetz, Chur

Building the new bridge

The canton consulted Maillart as an expert on the bridge's construction. An elderly Versell later told me that Maillart had liked his design. Once the contract was awarded, building the bridge took just six months. The northern abutment was founded on 68 driven wooden piles. At the end of May 1928, the concrete arch was poured on a centring structure held up by four fan-shaped supports standing in the river. On 10 June the Rhine flooded, washing away three of the four supports. Because seventeen days had already passed since the arch had been poured, the concrete survived the premature stripping of the formwork undamaged. The road deck was constructed accorded to plan and on 31 July, Roš carried out load tests on the new bridge (Fig. left page). He described the bridge as »a construction that is very well made in every respect«. The bridge was then opened to the public.

The load tests were especially interesting concerning the above-mentioned response of arch and road deck under unilateral loads. The tests showed that the road deck absorbs more of the bending moments than Versell assumed, with the road

mung auch auf die Fahrbahn und weckt in dieser einen Widerstand, der den Bogen zwar entlastet, in der Fahrbahn aber zu zusätzlichen Beanspruchungen führt. Versell berechnete, dass die Biegemomente sich im Verhältnis ¾ zu ¼ zwischen Bogen und Fahrbahn aufteilen würden.

Der Bau der neuen Brücke

Der Kanton zog Maillart als Experten für die konstruktive Durchbildung der Brücke hinzu. Maillart habe die Brücke gefallen, sagte mir Versell in hohem Alter. Der Bau der Brücke dauerte nach der Vergabe ganze sechs Monate. Das nördliche Bogenwiderlager wurde auf 68 gerammte Holzpfähle fundiert. Ende Mai 1928 betonierte man den Bogen auf einem Lehrgerüst, das sich auf vier fächerförmige Joche im Fluss abstützte. Am 10. Juni riss ein Hochwasser des Vorderrheins drei Joche des Lehrgerüsts weg. Zu diesem Zeitpunkt war der Bogen bereits 17 Tage lang betoniert und hielt dadurch der verfrühten Ausrüstung ohne Schaden stand. Die Erstellung des Überbaus erfolgte plangemäß, und schließlich führte Roš am 31. Juli die Belastungsversuche an der neuen Brücke durch (Bild Seite 102). Er bezeichnete die Brücke als »ein in jeder Beziehung sehr gut gelungenes Bauwerk«. Anschließend konnte die Brücke eröffnet werden.

Besonders interessant waren die Belastungsversuche hinsichtlich des oben erwähnten Zusammenwirkens von Bogen und Fahrbahnträger bei einseitigen Lasten. Die Versuche zeigten, dass sich die Fahrbahn stärker an der Aufnahme der Biegemomente beteiligte, als Versell angenommen hatte, je nach Laststellung übernahm die Fahrbahn mehr als die Hälfte der Momente. Obwohl Versell die massiven, außenliegenden Brüstungen in regelmäßigen Abständen durch Fugen getrennt hatte, besitzen diese offensichtlich dennoch eine versteifende Wirkung. »Die Vorderrheinbrücke Tavanasa entspricht einer außerordentlichen Anstrengung aller Beteiligten, von den Vertretern der Gemeinde und des Kantons über den Projektverfasser bis zu allen mit der Ausführung Verbundenen. Sie führt vor Augen, was Menschen in einer Notsituation zu leisten imstande sind. Von der Zerstörung der alten bis zur Belastungsprobe der neuen Brücke vergingen lediglich zehn Monate« schreibt ETH-Professor Peter Marti in seinem Gutachten zur Erhaltungswürdigkeit der Vorderrheinbrücke.

Würdigung der Vorderrheinbrücke

Versells Brücke mit ihrem äußerst schlanken Bogen ist einerseits eine virtuose technische Leistung, repräsentiert andererseits auch die stark von der Heimatschutzbewegung geprägte Brückenarchitektur der »gemäßigten Moderne«, die sich nicht scheute, dünne Betonbauteile mit kräftigem Natursteinmauerwerk zu verbinden, wenn dies aus der Aufgabe der jeweiligen Konstruktionsglieder begründet werden konnte. Diese Bauten sind nicht durch ihre Radikalität interessant, sondern durch ihren Versuch, materielle und stilistische Gegensätze durch geschickte Balance unterschiedlicher Bauteile und Baustoffe auszugleichen. Der Beton wird als wertvolles, nicht als Masse zu verwendendes Material aufgefasst. Bauteile, die statt betoniert gemauert werden können, werden in Stein errichtet.

Nach ihrer statischen Wirkungsweise bildet die Brücke eine Mischform zwischen eingespanntem Bogen und versteiftem Stabbogen. Dadurch gewinnt die Vorder-

deck absorbing more than half the load in some cases depending on load position. Although Versell divided the solid outer balustrades with joints at regular intervals, they clearly give added rigidity nonetheless.

»The bridge over the Rhine in Tavanasa embodies an extraordinary effort on the part of all involved,« wrote Prof. Peter Marti from the Swiss Federal Institute of Technology in Zurich (ETH) in his report on the bridge's eligibility for preservation status: »from the representatives of the local and canton authorities, to those who drew up the project, to everyone involved in the construction work itself. It shows what people are capable of in an emergency. From the destruction of the old bridge to the load testing of its replacement, just ten months passed.«

An appreciation of the Tavanasa Bridge

As well as being a technical masterpiece, Versell's bridge with its extremely slender arch also represents a »tempered modernism« in bridge-building that was strongly influenced by the cultural heritage movement and which didn't shy away from combining thin concrete components with substantial natural stone masonry if this appeared justified by the function of the element in question. These structures are interesting not just because of their radical approach, but also in their attempt to reconcile material and stylistic opposites with a skilled balance of different components and building materials.

Concrete is treated as a valuable material, not something to be used in mass. Parts of the structure that could be executed in masonry rather than concrete were realized in stone.

In structural engineering terms, the bridge is a hybrid of fixed arch and stiffened arch. As such, the Tavanasa bridge is conceptually related to Christian Menn's later bridges (Tamins, Viamala, Ponte Nanin, right picture on page 97).

In his lectures, Menn pointed out that even in the case of slender concrete bridges, it makes sense, in addition to the road deck, to use the »natural prestressing« in an arch to absorb part of the bending moments. Instead of a radical division of labour between an arch subject purely to pressure and a road deck subject only to bending (as in Maillart's arch), he used a more complex structural system with a carefully dosed distribution of the bending load. In his design for Tavanasa, Versell anticipated such ideas.

The Rhine bridge in Tavansa is the most important bridge built by the civil engineer Walter Versell (1891-1989). He represents the type of the 19th century engineer, capable of working on a broad range of tasks. This corresponded to the situation in a mountainous region with comparatively few firms of civil engineers. Bridge-building was only one aspect of his activities, but one in which he achieved great things: spanning 86 metres, his Ruseinertobel Bridge was one of the biggest arch bridges in Grisons when it was built in 1938.

Other bridges by Versell that are worth mentioning are the Reno di Lei arch bridge in Avers (1960), the demolished Rhine bridge in Felsberg (1935, concrete beam bridge) and the Glenner Bridge in Surin (1937, arch bridge).

At the same time, Versell worked in hydraulic engineering, was secretary of the Grisons-St. Gallen Rhine Association and later water management engineer for the Canton of Grisons.

rheinbrücke eine konzeptionelle Verwandtschaft zu Christian Menns späteren Scheiben-Platten-Bogenbrücken (Rheinbrücke Tamins, Viamalabrücke, Ponte Nanin ... Bild Ponte Nanin auf Seite 97, rechts). Menn wies in seinen Vorlesungen darauf hin, dass es auch bei schlanken Betonbrücken durchaus sinnvoll sei, neben dem Fahrbahnträger die »natürliche Vorspannung« in einem Bogen zur Aufnahme der Biegemomente mit zu benutzen. An die Stelle einer radikalen Arbeitsteilung zwischen rein gedrücktem Bogen und rein auf Biegung beanspruchter Fahrbahn (wie in Maillarts Stabbogen) trat ein komplexeres statisches System mit vom Projektverfasser dosierter Aufteilung der Biegebeanspruchung. Versell nahm bei der Brücke von Tavanasa derartige Überlegungen vorweg.

Die Vorderrheinbrücke Tavansa ist die bedeutendste Brücke des Ingenieurs Walter Versell. Walter Versell (1891-1989) vertritt den Ingenieurtyp des 19. Jahrhunderts, der eine breite Palette von Aufgaben bearbeiten kann. Dies entspricht seiner Situation in einer Bergregion mit vergleichsweise wenigen Ingenieurbüros. Brückenbau war nur eine Seite seiner Tätigkeit, aber eine mit bedeutenden Leistungen: seine Ruseinertobelbrücke (1938) gehörte bei ihrem Bau mit 86 Metern Spannweite zu den größten Bogenbrücken Graubündens.
Weitere erwähnenswerte Brücken von Versell sind die Bogenbrücke Reno di Lei im Avers (1960), die abgebrochene Rheinbrücke Felsberg (1935, betonierte Balkenbrücke) und die Glennerbrücke Surin (1937, Bogenbrücke). Daneben engagierte sich Versell im Wasserbau, war Sekretär des Rheinverbands Graubünden-St. Gallen und später Wasserwirtschaftsingenieur des Kantons Graubünden. Nach Aussage seiner Tochter löste Versell den verhärteten Konflikt um den Bau des Großstausees Splügen-Rheinwald durch die Ausarbeitung des Projekts für den Lago di Lei-Stausee. Bereits in den 1930er-Jahren befassten sich Versell und Kantonsingenieur Abraham Schmid in ihrer Freizeit mit der Idee des San Bernardino-Straßentunnels, die dann mit bemerkenswerter Zähigkeit in den folgenden Jahrzehnten zum konkreten Projekt reifte.
Zu erwähnen ist Versells Zusammenarbeit mit dem Churer Architekten Walther Sulser (1890-1983), die in der Markthalle Chur (1934) und der Erlöserkirche Chur (1937) ihren Höhepunkt fand. Beides sind Bauten der wirtschaftlichen Krisenzeit, auf äußerste Sparsamkeit bedacht, dafür mit einer intensiven Entwurfsarbeit verbunden: Der Turm der Erlöserkirche besitzt Betonwände von nur 12 Zentimetern Stärke.

Die Vorderrheinbrücke Tavanasa steht an einem brückengeschichtlich bedeutenden Ort, zeigt einen ungewöhnlich schlanken Betonbogen, der von gemauerten Pfeilern und Widerlagern wirkungsvoll eingerahmt wird, ist ein Hauptwerk des bedeutenden Ingenieurs Versell und lässt sich in der Entwicklung des Betonbrückenbaus präzis zwischen Maillart und Menn platzieren.

According to his daughter, Versell resolved a deadlocked conflict over the building of the Splügen-Rheinwald reservoir by developing a project for the Lago di Lei reservoir. In the 1930s, Versell and cantonal engineer Abraham Schmid worked in their own time on the idea of the San Bernardino road tunnel, helping it to become a concrete project over the following decades with remarkable tenacity.

It is also worth mentioning Versell's cooperation with the Chur architect Walther Sulser (1890-1983) that culminated in the market hall (1934) and the Church of the Redeemer (1937), both in Chur. They were built in a period of economic crisis, combining the utmost economy with intensive work on the design: the spire of the church has concrete walls just 12 centimetres thick.

The bridge over the Rhine in Tavanasa stands at a significant location in the history of bridge building, with an unusually slender concrete arch that is strikingly framed by the masonry of the pillars and abutments. It is a major work by the important civil engineer Versell and occupies a position in the development of concrete bridge construction between Maillart and Menn.

Blick auf Tavanasa und Danis,
Beginn der Instandsetzung, April 2017

View of Tavanasa and Danis, start of
renovation work, April 2017

Instandsetzung | Andrin Herwig: Die Versell-Brücke wurde instandgesetzt

Die in die Jahre gekommene Versell-Brücke hat ausreichende Tragsicherheits-Reserven, zumal sie in Zukunft nicht mehr von Autos, Bussen und Lastwagen, sondern nur noch von Fußgängern und Radfahrern genutzt wird. Teure Eingriffe in die Fundationen waren nicht nötig. Die Hauptursache der Schäden – eindringendes chloridhaltiges Wasser – konnte durch eine neue, dauerhafte Abdichtung und eine kontrollierte Entwässerung eliminiert werden. Somit kann bei der instandgesetzten Brücke mit geringen Unterhaltskosten und einer langen Lebensdauer gerechnet werden.

Ausgangslage

Die Brücke wurde zur Bauzeit und auch danach im Auftrag des Tiefbauamts gut dokumentiert; es gibt detaillierte Zustandsaufnahmen und Nachrechnungen der Brücke für Straßenverkehrslasten (Bänziger Partner, Chur, 2005) und eine Vermessung der Brücke (Schneider, Chur). Dabei wurde auch die Form des Bogens exakt aufgemessen. Weiter wurden Belastungsversuche, Messungen des Chloridgehalts und Untersuchungen zur Gefahr einer Alkaliaggregatreaktion durchgeführt.
Die Brücke war in schadhaftem, teilweise auch schlechtem Zustand. Dafür waren vor allem die 1978 an der Fahrbahn angebrachten Entwässerungsröhrchen verantwortlich – seither konnte abfließendes (oft tausalzgesättigtes) Wasser auf den Bogen tropfen und hat dort Salz- und Frostschäden verursacht. Auch die Fahrbahnträger und die Untersicht der Konsolen wurden von Salzwasser benetzt. Korrodierte Bewehrungen und Betonabplatzungen waren die Folge. Messungen zeigten eine Konzentration des Salzgehalts vor allem an den immer wieder durchfeuchteten Stellen, während trockenere Partien unschädliche Chloridgehalte aufweisen.

The ageing Rhine bridge has sufficient reserves of structural safety, especially since it will be used in future not by cars, busses and trucks, but only by pedestrians and cyclists. Costly work on the foundations was not necessary. The main causes of damage – penetration by water with high chloride levels – were eliminated by a new, permanent seal and controlled drainage. As a result, the restored bridge will have low maintenance costs and a long life.

Initial situation

At the time of its construction and in the intervening decades, the bridge was well documented at the behest of the Civil Engineering Department, with detailed evaluations, recalculations of traffic load (Bänziger Partner, Chur, 2005) and a survey (Schneider, Chur) during which the form of the arch was accurately measured. In addition, load tests were carried out, as well as measurements of chloride levels and assessments of the risk of an alkali-aggregate reaction. The bridge was damaged, in some areas badly. This poor condition was due mainly to the drainage tubes installed on the road deck in 1978 – since then, water flowing off (often saturated with de-icing salt) dripped onto the arch, where it caused salt and frost damage. The deck supports and the underside of the deck panels were also affected by salt water, leading to corroded reinforcement bars and spalling. Measurements showed a particularly high concentration of salt in places that were repeatedly soaked, while drier parts showed harmless levels. There were no indications of alkali-aggregate reactions; the unevenness observed during levelling was caused by creep and shrinkage of the concrete, and by temperature-related expansions of the bridge.

Bestandsaufnahme der Schäden am Brückenbogen, April 2017

Surveying damage to the bridge arch, April 2017

Die Lunetten der Brüstung sind 1978 halb zubetoniert worden. Eine Bewegungsfuge der Brüstung in originalem (Mitte) und in beschädigtem Zustand (rechts), mit den Resten eines Reparaturversuchs. Bestandsaufnahme April 2017

In 1978, the openings in the balustrade were half filled in with concrete. An expansion joint on the balustrade in original (centre) and in damaged condition (right), with the remains of an attempted repair. Site survey, April 2017

Es gab keine Hinweise auf Alkaliaggregatreaktionen; die durch Nivellementmessungen beobachteten Höhenbewegungen wurden durch Schwinden und Kriechen des Betons sowie durch temperaturbedingte Ausdehnungen der Brücke verursacht. Die detaillierte statische Überprüfung durch das Büro Conzett Bronzini Partner, Chur zeigte, dass die Brücke für die künftige Nutzung durch Langsamverkehr genügende Tragreserven aufweist. Damit konnten auch größere korrosionsbedingte Verluste des Bewehrungsquerschnitts akzeptiert werden.

Die Überprüfung erfolgte an einem ebenen Stabmodell, welches die effektive Bogengeometrie beinhaltete. Die Abweichungen der Bogenachse von der theoretischen Form betrug weniger als 5 Zentimeter. Die in die Rechnung eingeführte Nutzlast bestand aus einem Menschengedränge (400 kg/m^2). Die Last wurde an den ungünstigsten Orten platziert. Für die Beanspruchungen des Bogens wurden die Effekte zweiter Ordnung berücksichtigt. Dabei wurde die Rissbildung in den Querschnitten modelliert und ein entsprechend tiefer Elastizitätsmodul für sämtliche Querschnitte eingesetzt. Die Resultate der statischen Überprüfungen zeigten für den Bogen große Trag-reserven. Hingegen sind die Auswirkungen der versteifenden Fahrbahn auf deren Schnittgrößen beträchtlich. Zwar sind die meisten Querschnittswiderstände auch hier mit der neuen Nutzung nicht ausgeschöpft; nur bei den inneren Feldern der Fahrbahn nahe dem Bogenscheitel überschreitet das Bemessungsmoment den gerechneten Tragwiderstand um elf Prozent. Bestärkt wird dieses Resultat dadurch, dass genau an dieser Stelle die größten Schäden am bestehenden Tragwerk aufgetreten sind. Allerdings wurden diese auch durch den Chlorideintrag eines defekten

The detailed structural analysis by Conzett Bronzini Partner, Chur, showed that the bridge has sufficient load-bearing strength for its future usage by pedestrians and cyclists. This also meant it was possible to accept a significant corrosion-related reduction in the effective area of reinforcement.

The analysis was carried out using a two-dimensional bar model including the effective geometry of the arch. The deviation of the axis of the arch from the theoretical form was less than 5 centimetres. The live load used in the calculation consisted of a crowd of people (400 kg/m^2). The load was placed at the least favourable position. For the loads on the arch, second-order effects were taken into account. Crack formation was factored in and a corresponding lower elasticity module was used for all cross-sections.

The results of the structural analysis showed that the arch has major reserves of load-bearing strength. By contrast, the effects of the stiffening road deck on its internal forces are considerable: although most of the section moduli are not exhausted by the new usage, on the innermost fields of the road deck near the apex of the arch, the design forces exceed the calculated load-bearing capacity by eleven percent. Moreover, this is precisely where the worst damage to the existing load-bearing structure is found, aggravated by chloride channelled in by a defective drainage tube. The bridge is structurally sound nonetheless, because there are sufficient plastic reserves to transfer the bending moments and because the strength of the reinforcements was deliberately underestimated in the calculations. To be on the safe side, however, the reinforcements were also locally strengthened on the western edge of the underside of the arch.

Bewehrungseisen wurden freigelegt, wo sich die Betonüberdeckung gelockert hatte, und mit einem Schutzanstrich versehen, Juli 2017

Where the concrete covering them had come loose, reinforcing bars are laid bare and given a protective coating, July 2017

Bestandsaufnahme der Schäden mit Dzevad Jasarevic (J. Erni AG) und Andrin Herwig (Conzett Bronzini Partner), April 2017

Surveying damage with Dzevad Jasarevic (J. Erni AG) and Andrin Herwig (Conzett Bronzini Partner), April 2017

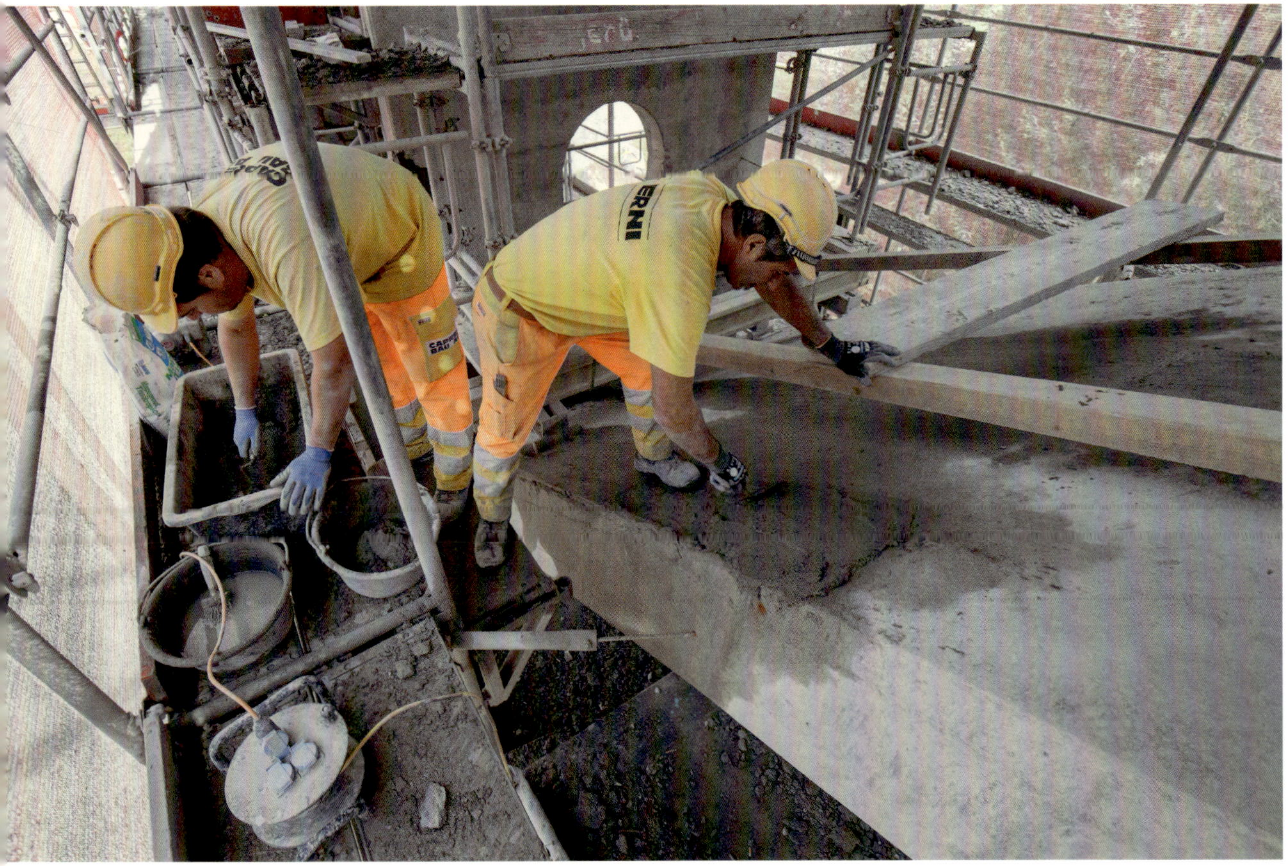

Entwässerungsröhrchens gefördert. Die Tragsicherheit des Bauwerks ist trotzdem gewährleistet, da einerseits genügend plastische Reserven für eine Umlagerung der Biegemomente vorhanden sind und bei der Überprüfung ein eher tief angesetzter Widerstand für die Bewehrung angenommen wurde. Sicherheitshalber wurde die Bewehrung dennoch am westlichen Rand der Bogenunterseite lokal verstärkt.

Der Baugrund besteht auf der Südseite aus mit Felsblöcken durchsetztem Kiessand. Auf der Nordseite steht das Bogenwiderlager in lehmigem Sand. Deshalb wurde dort auf Pfählen gegründet. Seit dem Bau der Brücke sind keine Setzungen der Fundationen festgestellt worden. Demzufolge konnte auf Verstärkung der Fundationen verzichtet werden – zumal die zukünftigen Verkehrslasten deutlich geringer ausfallen.

Um die Brücke für die nächsten Jahrzehnte zu ertüchtigen, ging es bei der Instandsetzung vor allem darum, das Abfließen des Wassers zu kontrollieren – dies erfolgte mit einer neuen Abdichtung der Fahrbahn und sorgfältig geplanter Entwässerung.

Die Brüstungen wiesen nur wenige Schäden auf. Während der Arbeiten wurde nach Abbruch des 18 Zentimeter starken Asphaltbelags eine intakte Polymerbitumenabdichtung angetroffen, die 1978 eingebaut worden war. Dadurch war der Beton der Fahrbahnplatte, zwischen den Längsträgern auch an deren Unterseite, praktisch unversehrt. Dies konnte als Hinweis darauf gesehen werden, dass vor dem Einbau der Polymerbitumenabdichtung kein Tausalz verwendet worden ist.

Die durch die schlecht funktionierende Entwässerung und Tausalzeintrag verursachte Korrosion der Bewehrungen mit Abplatzungen an der Fahrbahnplatte, direkt unterhalb der Brüstung, an den Außenseiten der Längsträger und am Bogen war jedoch teilweise gravierend.

On the south side, the ground consists of gravelly sand containing rocks. On the north side, it consists of clay-bearing sand, so the abutment here was built on piles. Since the bridge was built, no settling of its foundations has been registered. As a result, there was no need to reinforce the foundations – especially since future traffic loads will be considerably smaller. To fit the bridge for the decades to come, the restoration plan focussed on controlling the flow and discharge of rain water. This was achieved by resealing the road deck and installing a carefully planned drainage system. The balustrades were largely undamaged. During work, after the removal of an 18 centimetres layer of asphalt, an intact polymer bitumen seal applied in 1978 came to light. As a result, the concrete of the road deck, including the underside between the longitudinal beams, was essentially intact. This was taken as proof that de-icing salt had not been in use prior to the application of the polymer bitumen seal. By contrast, the damage to reinforcement bars caused by faulty drainage and corrosion by salt water, and the resulting spalling on the road deck, were severe in places: directly below the balustrade, on the outer surfaces of the longitudinal beams, and on the arch.

Restoration

Shortly after completion of the new road bridge, it was possible to begin restoration work on the old bridge. Where spalling had occurred, the rebars were given a rustproof coating and re-covered with fibre-reinforced concrete, with the new surface replicating the appearance of the original formwork. This treatment was limited to damaged areas only; »preventive« repairs were not carried out. All surfaces exposed to moisture were protected by hydrophobization. This minimal approach ensured that as much original substance as possible was retained, allowing the available funds to be spent mainly on key improvements.

Since rainwater drains off both sides through closely spaced spouts, the downpipes installed on the abutment pillars in the 1970s could be removed. Removing the pavement, which had been added on top of the road deck, revealed the full size of the semi-circular openings in the balustrade that had been partly blocked.

Reparatur der Abplatzungen auf der Oberseite des Bogens (links), September 2017
Als Grundlage für den ultrahochfesten Faserbeton (UHFB) wird eine Mörtelschicht mit seitlichem Gefälle aufgebracht, Juli 2017

Repairing spalling on the upper surface of the arch (left), September 2017
Mortar is applied with a cambered profile to support the ultra-high-performance seal (UHPFRC), July 2017

Instandsetzung

Kurz nach Fertigstellung der neuen Straßenbrücke konnte mit den Instandsetzungsarbeiten bei der alten Versell-Brücke begonnen werden. Die Bewehrung schadhafter Stellen wurde mit einer Rostschutzbeschichtung versehen und mit einer kunststofffaserverstärkten Mörtelschicht neu überdeckt. Dabei beschränkte man sich auf eine Instandsetzung beschädigter Partien und führte keine »präventiven« Reparaturen durch. In der neuen Oberfläche wurde das ehemalige Schalungsbild nachgebildet. Sämtliche feuchte-exponierten Oberflächen wurden durch Hydrophobierung geschützt. Mit dieser minimalen Art der Instandsetzung gelang es, so viel Originalsubstanz wie möglich zu erhalten und das Geld vorwiegend für substantielle Verbesserungen einzusetzen.
Da die Ableitung des Regenwassers beidseitig über mehrere, fein verteilte Speier erfolgt, konnten die in den 1970er-Jahren angebrachten Fallrohre an den Kämpferpfeilern wieder entfernt werden. Nach Abbruch des Trottoirs, das der Fahrbahnplatte aufgedoppelt war, zeigen sich die zuvor teilweise geschlossenen halbkreisförmigen Brüstungsöffnungen wieder in der ursprünglichen Größe.

Der Betonabtrag auf der Fahrbahnplatte beschränkte sich auf leichte Anpassungen der Querschnittsgeometrie und auf das Aufrauhen der Oberfläche. Auf die aufgeraute Oberseite der Fahrbahnplatte wurde eine Ausgleichsschicht mit dachförmigem Gefälle (Bild rechts) aus Mörtel aufgetragen, deren Oberfläche wiederum mit Wasserhochdruck für die Aufnahme der Abdichtung in Form eines ultrahochfesten Faserfließbetons (UHFB) vorbereitet wurde. Die Entwässerung der Brückenfahrbahn erfolgt über das Längsgefälle der Brücke. Ein lokales Gegengefälle entlang des Fahrbahnrandes leitet das Wasser von den Brüstungen weg.
Der Zustand der Widerlager und Kämpfer war zum größten Teil gut. Das Natursteinmauerwerk wurde oberflächlich gereinigt. Die wenigen offenen Fugen der natursteinverkleideten Kämpferpfeiler wurden neu vermörtelt. Der Aufbau des UHFB-Belags besteht aus zwei Schichten: die untere Schicht ist 2,5 Zentimeter stark, die obere 0,5 Zentimeter. Die obere Schicht wurde nass in nass auf die untere eingebracht. Sie enthält im Gegensatz zur unteren keine Stahlfasern. Auf den nassen UHFB streute man angefeuchteten Kies und drückte ihn leicht an. Die Nachbehandlung durch Abdecken mit Plastikfolie erfolgte unmittelbar nach dem Einbau.
Der UHFB-Belag wurde, in Brückenlängsrichtung gesehen, in fünf Etappen eingebracht. Im Gegensatz zu anderen Abdichtungsarten wie Gussasphalt usw. entfallen beim UHFB besondere Randabschlüsse. Der UHFB verbindet sich monolithisch mit dem Untergrund, das heißt, er ist kraftschlüssig und wasserdicht mit dem Beton verbunden. Die Details bei den Entwässerungen und den Dilatationen bestehen weitgehend auch aus UHFB und wurden im gleichen Arbeitsgang mit der Fahrbahnplatte gegossen (Bilder rechts).
Die Festigkeit des UHFB ist etwa fünf mal größer als diejenige eines normalen Betons. Erreicht wird dies durch den hohen Zementgehalt, eine hohe Konzentration an Stahlfasern und einen geringen Wassergehalt. Die Verarbeitbarkeit dieses selbstverdichtenden Betons wird durch Zugabe eines Hochleistungsverflüssigers verbessert. Sein niedriges Schwindmaß und seine hohe Zugfestigkeit gewährleisten Rissefreiheit und ein verbessertes Tragverhalten der Fahrbahnplatte.

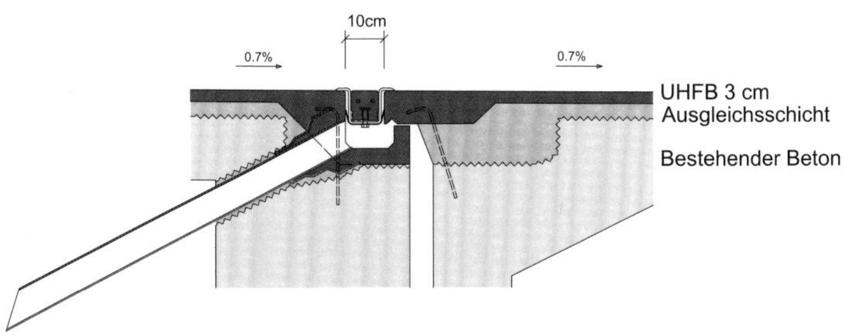

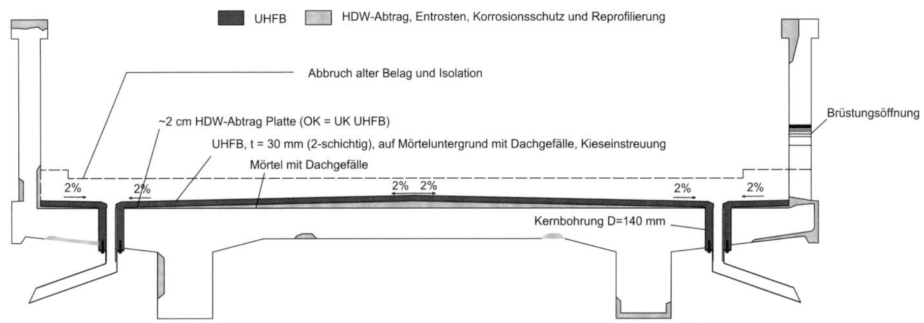

Removal of concrete from the road deck was limited to minor adjustments in cross-sectional geometry and a roughening of the surface. Onto this roughened surface, mortar was applied in a cambered profile (Fig. above, right), the surface of which was then water-blasted in preparation for the application of a seal consisting of ultra-high-performance fibre-reinforced concrete (UHPFRC). The road surface drains along the bridge's longitudinal gradient. Gutters along the length of the deck channel water away from the balustrades. The state of the abutments and pillars was mostly good. The surfaces of the stone masonry were cleaned. The few open joints on the stone-clad abutment pillars were repointed.

The UHPFRC seal consists of two layers: the bottom layer is 2.5 centimetres thick, the top 0.5 centimetres. The top layer was applied to the bottom layer wet on wet. Unlike the bottom layer, the top layer contains no steel fibres. Wet gravel was scattered onto the wet UHPFRC and lightly pressed down. Plastic sheeting was applied immediately after pouring.

The UHPFRC seal was poured in five sections, along the length of the bridge. Unlike other kinds of seal such as poured asphalt, UHPFRC does not require special edging. The UHPFRC bonds monolithically with the substrate, meaning the connection with the concrete is firm and waterproof. The details of the drainage system and expansion joints also consist largely of UHPFRC, poured in the same work phase as the road deck (Figs. above).

UHPFRC is roughly five times stronger than normal concrete. This is due to its high cement content, high concentration of steel fibres and small water content. The workability of this self-compacting concrete is improved by adding a superplasticizer. Its low degree of shrinkage and its high tensile strength ensure freedom from cracks and improved load-bearing properties of the road deck.

Bild links: Ausbildung des UHFB-Belags bei den Dilatationsfugen über den Kämpferpfeilern. Rinne mit Entwässerungsrohr in der Brückenachse
Picture left: Pouring the UHPFRC seal on the expansion joints over the abutment pillars. Channel with drainage pipe in the centre line of the cross section

Bild rechts: Gefälleausbildung mit Mörtel, UHFB-Belag, Entwässerung und Instandsetzung an der Fahrbahnplatte mit Brüstung
Picture right: Mortared camber, UHPFRC seal, drainage and restoration at the edge of the road deck with balustrade

gerüstbau-tscharner.ch

Damit man für die Instandsetzungsarbeiten überall an die Brücke herankommen kann, wird sie eingerüstet. Das ist kein Job für Angsthasen. Man sollte schwindelfrei sein und zupacken können, März 2017
To give access to all parts of the bridge during renovation work, scaffolding is erected. This is no job for the faint of heart. You need a head for heights and a firm grip, March 2017

gerüstbau-tscharner

Tel. 081 650 02 60

Luzi

Die nicht originale, oberste Schicht der Fahrbahn wird abgetragen, April 2017
The upper layer of the road deck (not part of the original bridge) is removed, April 2017

Die 1978 – nicht nur für die seitlichen Fußwege – aufgebrachte Betonschicht wird mit dem Presslufthammer mechanisch abgespitzt
The layer of concrete added in 1978 – not just for the pavements – is removed with a pneumatic drill

Die freigelegte, ursprüngliche Fahrbahn wird mit dem Hochdruckreiniger aufgeraut, damit der Mörtel haften kann, Mai 2017
The uncovered original road deck is roughened by water-blasting to ensure that the mortar adheres properly, May 2017

gerüstbau-tscharner.ch
gerüstbau-tscharner.ch

Mit Hochdruck werden die Brüstungen gereinigt. Hinter dem Vorhang sieht das grafisch reizvoll aus, ist aber kein Kinderspiel. Mai 2017
The balustrades are cleaned by water-blasting. Seen through the sheeting this has a certain graphical appeal, but it's no easy task. May 2017

Der Hochdruckreiniger hinterlässt Spuren, die an Spiralfedern erinnern. Das knallige Rot des Vorhangs ist längst verblasst. Juli 2017
The water-blaster leaves marks that recall springs. The bright red of the sheeting has long-since faded. July 2017

22.11.2017: Das Gerüst hätte längst abgebaut sein sollen. Aber beim Bauen gibt es immer Verzögerungen. Wenigstens das Licht ist installiert (Seite 140). Kommen Sie doch nach Tavanasa und schauen selbst!

22.11.2017: The scaffolding should have been dismantled by now. But with any construction project, there are always delays. At least the lights have been installed (page 140). Come to Tavanasa and see for yourself!

Fundaziun Nossa Punt | Sep Cathomas: Gründung und Stiftungszweck

Die Versell-Brücke ist im Inventar der historischen Verkehrswege der Schweiz aufgeführt. Dass die Standortgemeinde Brigels trotzdem nicht bereit war, die Brücke zu übernehmen, sorgte bei Anwohnern und in Fachkreisen für Unmut. Um den Erhalt des bedeutenden Objektes zu sichern, musste nach einem neuen Eigentümer gesucht werden. Dies hat die Denkmalpflege und insbesondere den Verein Nossa Punt stark beschäftigt. Weil der Verein die Instandsetzung sowie den künftigen Betrieb der Brücke aus Rechts- und Haftungsgründen nicht verantworten konnte, gründete dieser am 9. April 2016 die Stiftung Fundaziun Nossa Punt mit dem Zweck, die Versell-Brücke zu übernehmen, »sie als schützenswertes Denkmal zum Gebrauch für den Langsamverkehr wie eine gemeindeeigene Verbindungsstraße zu erhalten sowie sie gemäß den Grundsätzen der Denkmalpflege instand zu setzen und zu unterhalten«. Nach der grundbuchamtlichen Beurkundung und der Übernahme der Brücke am 30. September 2016 wurden die zukünftige Nutzung und die Betriebshaftung für die Brücke mit der Gemeinde Brigels vertraglich geregelt.

Die Beschaffung der finanziellen Mittel zur Deckung der Baukosten von etwa 900000 Franken war für den Stiftungsrat trotz des starken Rückhalts durch den Kanton, die Denkmalpflege, das Bundesamt für Straßen, sowie durch Stiftungen und Gönner eine große Herausforderung. Mit Respekt vor der alten Bausubstanz und mit fachlichem Können ist die Versell-Brücke inzwischen wieder ertüchtigt worden und bleibt als wertvolles Kulturgut und als Zeugnis hoher Ingenieur-Baukunst für zukünftige Generationen erhalten. Anerkennung und herzlichen Dank allen, die dazu beigetragen haben! Sep Cathomas, Präsident Fundaziun Nossa Punt

Sep Cathomas: Founding and Purpose

Versell's bridge in Tavanasa is on Switzerland's index of historical transport routes. In spite of this, the municipality of Brigels was unwilling to take on the bridge, much to the annoyance of experts and local residents. To secure the future of this important structure, a new owner had to be found, a task dealt with by the Office for Historic Monuments and especially the association »Nossa Punt«.
For reasons of law and liability, the association was not able to take responsibility for restoring and operating the bridge. For this reason, it created the »Fundaziun Nossa Punt« on 9 April 2016 to purchase it, »to preserve this valuable historical monument for use by pedestrians and cyclists as a connecting route owned by the municipality, and to restore and maintain it in keeping with the principles of heritage preservation.« Once the acquisition of the bridge had been entered in the land register on 30 September 2016, the future usage and operating liability for the bridge was agreed by contract with the municipality of Brigels.

Raising the 900,000 Swiss francs needed to cover the building costs was a challenge for the Foundation, in spite of solid support from the Canton, the Office for Historic Monuments, the Federal Office for Roads, and various donors and benefactors. With respect for the original structure and with specialist expertise, Versell's bridge now shines like new, preserved as a valuable cultural asset and engineering landmark for future generations. Compliments and heartfelt thanks to all those who helped make this possible! Sep Cathomas, President, Fundaziun Nossa Punt

Sep Cathomas: Constituziun ed intent

La punt da Versell ei inscretta egl »Inventari da las vias da communicaziun istoricas da la Svizra«. Tuttina ei la vischnaunca da Breil buca stada pronta da surprender la punt en agen possess. Quei ha irritau ton habitontas e habitonts da Tavanasa sco era tscherchels d'experts. Per saver mantener quei impurtont object ein la tgira da monuments cantunala e surtut l'uniun Nossa Punt sestentadas d'anflar ina sligiaziun e surtut ina nova possessura. L'uniun saveva ord motivs giuridics buca responsar la sanaziun sco era il menaschi futur dalla punt. Perquei ha ella constituiu ils 9 d'avrel 2016 la fundaziun Nossa Punt cun la finamira da surprender e da mantener il baghetg »sco monument degn da protecziun per diever dal traffic plaun sco ina via da colliaziun communala sco era da sanar e mantegnair ella tenor ils princips da la tgira da monuments«. Cun la registraziun el cudisch funsil ils 30 da settember 2016 ei la punt ida en possess dalla fundaziun. Il diever e la responsabladad per il manaschi ei vegnius reglaus cun la vischnaunca da Breil en in contract. Rimnar ils mieds finanzials per cuvierer ils cuosts da baghegiar da rodund 900'000 francs ei stau ina gronda sfida per il cussegl da fundaziun, malgrad ch'il cantun, la tgira da monuments, igl uffeci federal da vias, fundaziuns, fauturas e fauturs han dau ferm sustegn. Cun respect per la veglia substanza dalla construcziun e cun savida professiunala ei la punt da Versell vegnida reparada e sepresenta ussa da niev. Aschia resta ella mantenida per las generaziuns futuras sco bein cultural prezius e sco perdetga d'ina ovra d'inschignier d'aulta qualitad. Renconuschientscha e cordial engraziament a tut quellas e quels che ha contribuiu a sia realisaziun.
Sep Cathomas, president fundaziun Nossa Punt

31.8.2017 Vorstand des Vereins Nossa Punt: Marlis Caduff, Alex Liechti, Rita Cathomas-Bearth, Daniel Carigiet (Marc Cathomen fehlt)

31.8.2017 Executive committee of Nossa Punt association: M. Caduff, A. Liechti, R. Cathomas-Bearth, D. Carigiet (M. Cathomen is missing)

31.8.2017 Suprastonza dall'uniun Nossa Punt: Marlis Caduff, Alex Liechti, Rita Cathomas-Bearth, Daniel Carigiet (Marc Cathomen maunca)

Autoren | Authors

Wilfried Dechau, Dipl.-Ing., DGPh, *1944 in Lübeck. 1980-87 Redakteur, 1988-2004 db-Chefredakteur. Buchautor (u. a. ›architektur abbilden‹). Gründung des Fotografie-Preises ›architekturbild‹ (1995), des architekturbild e.v. (2003) und der Fotogalerie f75 (2010). Seit 2004 freier Fotograf.

Ursula Baus, Dr.-Ing., *1959 in Kaiserslautern. Kunstgeschichtsstudium in Saarbrücken, Architekturstudium in Stuttgart und Paris. Promotion Uni Stuttgart. 1989-2004 db-Redakteurin. 2004 Mitgründung von frei04-publizistik. Seit 2010 im wissenschaftlichen Kuratorium der IBA Basel 2020.

Peter Egloff , *1950 in Zürich. Studium der Volkskunde, Europ. Volksliteratur und Soziologie. Zahlreiche Auslandeinsätze für IKRK, UNO, OSZE, EU. Neun Jahre Redakteur bei Schweizer Radio DRS, sieben Jahre Programmleiter der Televisiun Rumantscha. Seit 2002 freier Publizist.

Robert Kruker, Dr. phil., *1946. Zuerst Buchhändler, dann Studium. Von 1977-84 in Forschung und Lehre sowie als Autor tätig. Anschließend beim Schweizer Radio. Seit 2006 freier Publizist, u. a. ›Surselva, Wandern im Westen Graubündens‹. Lebt in Dardin.

Arno Camenisch, *1978 in Tavanasa. Schriftsteller, studierte am Schweizerischen Literaturinstitut in Biel, wo er auch lebt. Die beiden Bände »Hinter dem Bahnhof« (2010) und »Ustrinkata« (2012) aus der »Bündner Trilogie« spielen in seinem Geburts- und Kindheitsort Tavanasa.

Rita Cathomas-Bearth, *1949. Übersetzerin, Projektleiterin, aktiv in verschiedenen Kulturinstitutionen, ehemaliges Mitglied von nationalen Kommissionen und Institutionen des Natur- und Heimatschutzes. Präsidentin der Uniun Nossa Punt.

Simon Berger, *1981. Studium der Kunstgeschichte mit Denkmalpflege und Monumentenmanagement, seit 2014 Denkmalpfleger des Kantons Graubünden. Wohnt in Rhäzüns.

Jürg Conzett, dipl. Bauingenieur ETH/SIA, *1956. Bauingenieurstudium an der EPF Lausanne und an der ETH Zürich. Diplom 1980. Mitarbeit im Büro Peter Zumthor. Seit 1988 eigenes Ingenieurbüro (Conzett Bronzini Partner) in Chur. Wichtige Projekte: Punt da Suransuns, Traversinersteg.

Andrin Herwig, Dr. dipl. Bauingenieur ETH/SIA, *1969. Bauingenieurstudium an der ETH Zürich und EPF Lausanne. Diplom 1996. Mitarbeit in Planungsbüros. 2008 Promotion. 2008-14 Forschung und Entwicklung an der EMPA, Dübendorf. Seit 2014 Ingenieurbüro Conzett Bronzini Partner, Chur.

Wilfried Dechau, Dipl.-Ing., DGPh (born 1944 in Lübeck). Editor and editor-in-chief (1980-2004) of db. Author (incl. »architektur abbilden«). Founder of the architectural photography prize (1995), architekturbild e.v. (2003) and the gallery f75 (2010). Freelance photographer since 2004.

Dr. Ursula Baus (born 1959 in Kaiserslautern). Studied art history in Saarbrücken and architecture in Stuttgart and Paris. Doctorate Stuttgart University. 1989-2004: editor at db. 2004: co-founded frei04-publizistik. Since 2010 member of the Board of Trustees of IBA Basel 2020.

Peter Egloff (born 1950 in Zurich). Studied ethnology, European folk literature and sociology. Many foreign missions for ICRC, UN, OSCE, EU. Nine years as editor on Radio DRS, seven years as programming director for Televisiun Rumantscha. Since 2002 freelance writer.

Robert Kruker, Dr. phil. (born 1946). Worked as a bookseller before studying. 1977-84: researcher, teacher and writer, followed by work on Swiss radio. Since 2006 freelance writer, including »Surselva, Wandern im Westen Graubündens«. Lives in Dardin.

Arno Camenisch (born 1978 in Tavanasa). Writer, studied at the Swiss Institute of Literature in Biel, where he lives. The two volumes »Hinter dem Bahnhof« (2010) and »Ustrinkata« (2012) from his »Bündner Trilogie« are set in Tavanasa, where he was born and grew up.

Rita Cathomas-Bearth (born 1949 in Rabius). Translator and project manager, active in various cultural institutions, former member of the national commissions and institutions of the Offices for the Protection of Nature and Heritage. President of the Nossa Punt association.

Simon Berger (born 1981). Studied art history, heritage preservation and monument management. Since 2014 curator of monuments, Canton of Grisons. Lives in Rhäzüns.

Jürg Conzett, dipl. Bauingenieur ETH/SIA (born 1956). Studied civil engineering at EPF Lausanne and ETH Zurich. Diploma in 1980. Worked with Peter Zumthor. Since 1988: own engineering office in Chur (Conzett Bronzini Partner). Major projects: Punt da Suransuns, Traversiner Footbridge.

Andrin Herwig, Dr. dipl. Bauingenieur ETH/SIA (born 1969). Studied civil engineering at ETH Zurich and EPF Lausanne. Graduated in 1996. Worked in planning offices. 2008: doctorate. 2008-14: research and development at EMPA in Dübendorf. Since 2014 Conzett Bronzini Partner in Chur.